The Patrick Moore Practical Astronomy Series

Series Editor

Gerald R. Hubbell
Mark Slade Remote Observatory, Locust Grove, VA, USA

More information about this series at http://www.springer.com/series/3192

Radio and Radar Astronomy Projects for Beginners

Second Edition

Steven Arnold

 Springer

Steven Arnold
Mansfield, UK

ISSN 1431-9756   ISSN 2197-6562   (electronic)
The Patrick Moore Practical Astronomy Series
ISBN 978-3-030-54905-3   ISBN 978-3-030-54906-0   (eBook)
https://doi.org/10.1007/978-3-030-54906-0

Dedicated to
Marjorie, Kate, Minnie, and Alfie

Preface

This second edition book is divided into three parts.

Part I: The History and Science of Radio Astronomy

This section now covers in much more detail the history and science of radio astronomy. It also introduces you to how things work, including receivers, antennas, and feed cables, etc. Also included are updates on the latest changes and advancements made in the field of radio astronomy.

Part II: Self-Build Tried and Tested Projects

These projects were covered in the first edition but now contain updates and extra information that the reader will find useful when choosing what equipment will be suitable for different observing sites.

Part III: Radio Astronomy with a Software-Defined Radio (SDR)

This is entirely a new material showing how technology has progressed. It introduces the latest tools available to the amateur radio astronomer and walks readers through any pitfalls they may encounter setting up and running successful SDR equipment. Also included is advice on purchasing equipment, what software to use, where to get it, and how to set it up and use it. You will find in this section some fun projects to build. These are aimed at the beginner and move up to more advanced ones for the seasoned observer.

Note that credit lines are provided for figures taken from other sources. All other figures and images are my own.

Safety and Legal Matters

Please read this material before continuing through the rest of the book.

It is impossible to think of every eventuality, but if a few common sense precautions are put into practice, then no harm should come from building and using these projects.

Electrical Safety

1. *Electricity can kill!* If in any doubt about mains/grid power supply, consult a qualified electrician.
2. Remember it can get damp outside in the evening and early morning with dew from the grass and condensation as things cool down. Therefore, it is not advisable to have mains/grid power outside on an extension lead, as there is a risk of electric shock. There is the added danger that someone may trip over the lead in the dark.
3. If a soldering iron is being used, make sure it is not within the reach of children or pets, and allow it at least 30 min to cool down after use before putting it away.
4. If mains/grid power is to be used instead of batteries, seek the advice of a qualified electrician if in any doubt.

5. Please follow any instructions supplied with the kits carefully, especially where polarity is involved. Some electronic components are polarity sensitive. As a rule of thumb: check twice, solder once.
6. Replace any blown fuse with one of equal value.
7. If there is a local thunderstorm, switch off any receivers/monitors and computers and unplug antennas.
8. Under no circumstances apply mains voltage to a coax feed cable.

Do-It-Yourself (DIY) Safety

1. Take the precaution of wearing eye protection and other safety protection as and when the need arises.
2. If power tools are to be used, refer to the manufacturer's instructions and follow any advice given. The same applies to hand tools and others.
3. When soldering, make sure there is adequate ventilation to remove fumes generated by the soldering process.
4. Solder now comes in lead-free varieties, but it is still a good practice to wash hands after use or before eating or drinking.
5. When working at heights, for example, off a mast, be sure to wear the correct fitting harness. If working off a ladder, make sure it is the right ladder for the height. Do not overreach, and have someone trustworthy steadying the ladder.
6. If metal antenna masts are used, make sure they are earthed to prevent damage if struck by lightning.
7. When fitting the feed cable to a mast, allow enough cable to put a "U" bend at the base of the mast. This serves two purposes:
    (a)   It will allow rain to drip off and not run along the cable.
    (b)   If the antenna is struck by lightning and travels along the cable, it is more likely to discharge to earth at the base of the "U" bend, therefore hopefully protecting equipment further along.

Using Equipment in the Countryside

1. Please respect "no entry" and "private land" signs. Get the landowner's permission first before crossing their land.
2. Each country has its own laws regarding personal safety and the level to which this can be upheld. Take simple precautions such as carrying a cell phone and letting someone trustworthy know the plan and an estimated time of return. But, remember to stick to this plan.

3. It is also a good idea to carry extra food and water, plus a basic first aid kit, insect repellent, etc.
4. In the event of a thunderstorm, do not be tempted to shelter under a tree. Trees are basically full of water and make a good lightning conductor. However, the safest place is inside a metal-bodied vehicle.

Software Downloads

1. When software is downloaded from the internet, whether freeware or licensepaid, please respect the terms and conditions of its use.
2. If it is freeware and there is a donate button, consider leaving a donation of a few dollars, as this can be used to make improvements to the future versions of the software.
3. With the ever-increasing threat from cyber-attacks, it is a good practice to keep any anti-malware or anti-virus software up to date and scan any downloaded software with the aforementioned anti-malware or anti-virus software before it is opened or used on your computer.
4. A good idea is to have any personal information, files, documents, and images backed up elsewhere, perhaps on a separate hard drive or online "cloud" servers.

Software-Defined Radios (SDRs)

1. Software-defined radios (SDRs) are powerful tools capable of covering a large number of frequencies. Please respect the laws of the country in which it is being used. For example in the UK, it is not against the law to receive radio broadcasts from commercial or amateurs, but it is against the law to broadcast unless you have passed a radio operator's test and have a license to do so.
2. Receiving frequencies used to handle sensitive information in most if not all countries is against the law for obvious reasons. If you alight on such a frequency, do not make any recordings and leave the frequency immediately.

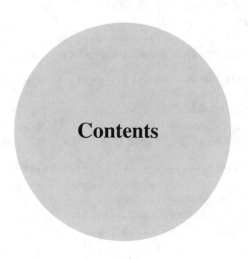

Contents

Part I

The History and Science of Radio Astronomy

Chapter 1

A Brief History of Radio Astronomy

First Attempts at Receiving Radio Transmissions from the Sun

In the late 1880s, a scientist by the name Heinrich Hertz conducted a series of experiments where he successfully transmitted and received radio waves across his laboratory. After doing this, Hertz noted in his journal, "I have successfully transmitted and received radio waves and have found them of no practical use." The unit of frequency, Hertz (Hz), would later be named in his honor.

The first documented attempt at receiving radio waves from the Sun was in the early 1890s by Thomas Edison. Edison was not only an innovator but also a brilliant salesman with a knack for convincing investors to believe in his ideas. Why Edison woke up one morning with the belief that the Sun emitted radio waves is unclear. Regardless, Edison sent one of his assistants to the Lick Observatory in California with instructions on how to construct an antenna and receiver. Unfortunately, there are no documents surviving with details of the construction and design plans for Edison's receiver and antenna. After this first failed attempt to receive radio waves from the Sun, Edison made no further attempts to repeat his experiment. At the time, he

---

The original version of this chapter was revised. The correction to this chapter is available at https://doi.org/10.1007/978-3-030-54906-0_20

was caught up in the great "Current War" with Nikola Tesla, wherein Edison had his direct current (DC) and Tesla had his alternating current (AC).

Around this time, radio waves were still new and not fully understood. The idea of wavelength and frequency was still years away. It was thought that radio waves could only travel in straight lines, like the beams of light coming from a lighthouse. No one was sure how they propagated, so a mysterious substance called "luminiferous aether" was invented to explain it. This aether was invisible but allowed radio waves to travel through it.

It wasn't until 1900, when Marconi demonstrated the first trans-Atlantic radio transmission from Poldhu in Cornwall to Newfoundland, that such long-distance radio transmission was proved possible. (It must be said that there is some conjecture concerning whether Marconi actually did make this radio transmission or not, and whether Marconi did in fact invent the radio. There is evidence that Nikola Tesla invented the radio, but Marconi beat him to the patent. We'll assume here that Marconi did make the transmission.)

After Marconi made this transmission, in a separate experiment, the idea that radio waves could only travel in straight lines was proven incorrect. Shortly thereafter, the belief in the aether, that mysterious medium that aided the propagation of electromagnetic waves, was finally abandoned.

Many theories were put forward how it was possible for radio waves to travel around the Earth. One theory suggested there could be a layer in the upper atmosphere that was able to reflect radio waves back down to the Earth and allow great distances to be covered. This theory turned out to be correct and will be covered in Chap. 4.

The first radio transmitters were nothing more than "spark coils." These work in a similar manner to the ignition coil on a petrol engine. A coil is charged up and when the electrical energy is released, a spark jumps across a gap between two electrodes. If this is fed to an antenna it makes a crude radio transmitter. There is no modulation to the signal—it is either on or off—but it is enough to send messages using Morse code.

Here is a link to a video showing a spark coil transmitter: https://www.youtube.com/watch?v=YSf93g0heUA.

Between the years of 1890 and 1897 a further attempt at receiving radio waves from the Sun was made by Sir Oliver J. Lodge. He built an antenna and receiver, but he too failed to receive anything. This attempt is not well documented, so it is not clear exactly why he failed. Maybe his equipment wasn't sensitive enough, he was trying the wrong frequency, or the solar activity was simply very low.

The next attempt to receive radio emissions from the Sun was made by two astrophysicists, J. Schiener and J. Wilsing between the years of 1897 and 1899. They constructed their own antenna and receiver and let their experiment run for just over a week, but they also failed to receive radio emissions from the Sun. From this, they incorrectly theorized that the atmosphere must therefore be absorbing all radio waves coming from the Sun. They were on the right track, as the Earth's atmosphere does absorb some radio waves.

These failed experiments added fuel to the theory of a reflective layer in the upper atmosphere that was capable of reflecting radio waves back towards the Earth and which in turn must also reflect any radio waves from the Sun back into space.

Between the years 1899 and 1901, and based on the incorrect theories developed from the week-long experiment, French graduate student Charles Norman took the interesting approach of taking his receiver and antenna up a mountain to an altitude of over 10,000 ft (3000 m). Norman thought that if Schiener and Wilsing were right about the atmosphere absorbing radio waves, he may be able to receive something at this altitude, as he was above the thickest part of the atmosphere. Yet he also failed to receive anything.

Experts looking back over his notes and equipment have agreed that Norman's equipment would have been sensitive enough to receive some radio emissions from the Sun. Unfortunately, he carried out the experiment at the time of solar minimum, and it was just a case of bad luck he didn't receive anything.

After all these early attempts ended in failure, the idea of receiving radio waves from the Sun went out of fashion, also in part due to the onset of the First World War. It wasn't until the middle of World War II that this mystery would finally be solved—more about this in later chapters.

First Attempts at Listening for Radio Signals from Space

The first attempts at listening for radio signals from space were carried out by Marconi and Tesla shortly after the invention of the radio in the early 1920s. Marconi fitted a radio receiver to a ship and sailed into the Atlantic Ocean to get away from human made interference. Tesla built a large radio transmitter in Colorado Springs. Tesla's device was capable of both receiving as well as transmitting signals. Both men heard strange whistling and chirping noises, and both failed to realize they were produced by the Earth itself within its own magnetic field.

The whistling noises are now known as "whistlers." These are very low frequency (VLF) signals from lightning discharges that have traveled up into the ionosphere and interacted with the Earth's magnetic field lines. The chirping noises are now known as "tweeks." These too are VLF signals. Tweeks are created by charged particles from the solar wind interacting with the Earth's magnetic field. VLF signals are strange-sounding signals.

It is no surprise that when they were first heard, it wasn't known whether they came from space or had a more paranormal nature.

From this experience, Tesla became obsessed and went on to build a radio receiver that came to be known as his "spirit radio." When using his spirit radio Tesla heard even stranger sounds that he thought sounded like voices, which led him to believe he was picking up spirit voices of the dead. The Fig. 1.1 below is a modern version of Tesla's spirit radio.

Edison tried looking for spirit voices within the amplified white noise of his phonograph. When this failed, he theorized building a device to communicate with the recently departed and record their spirit voices for relatives to hear. This device came to be known as the "telephone to the dead,"

Fig. 1.1 A modern copy of a Tesla spirit radio

or the "spirit telephone." There is no record of this device being found or even if it was built.

Anyone interested in building a Tesla spirit radio please see Chap. 19.

The Discovery of the Ionosphere

In the early to mid-1920s, extensive work was carried out by a number of scientists and researchers to prove the existence of a reflective layer high up in the Earth's atmosphere. Over several years, radio signals were bounced off this reflective layer and the returning signals were studied in order to try and understand its properties.

This reflective part of the atmosphere, later to be known as the ionosphere, was found to reflect certain frequencies of radio waves like a mirror reflecting a beam of light. It was first thought that no radio waves could travel through the ionosphere. As the technology improved, however, it was demonstrated that the ionosphere would indeed allow certain radio waves to pass through it, so long as they were at the right frequency.

A practice called "sounding the ionosphere" was then introduced. This involved reflecting radar beams off the ionosphere from the Earth's surface to study its properties and understand its effects at different frequencies. Other attempts to understand the ionosphere's properties included sending transmitters high into the Earth's atmosphere using high altitude balloons. Even rockets have been used, but this took place later, after reliable rocket technology had been developed.

Ionospheric studies and testing are carried on up to the present day, analyzing changes to the structure of the ionosphere and any changes due to seasonal variation. Today, satellites are employed to help perform tests involving the ionosphere.

In 1927 a scientist named Sydney Chapman proposed the first mathematical model of the Earth's ionosphere. This was considered at the time to be the best model to explain the then-known properties of the ionosphere and took into account how the ionosphere changes throughout the year.

The Birth of Radio Astronomy

In the early 1910s, the fastest way to send communication across the Atlantic Ocean was to use telegraph. The wires for telegraph equipment had to run along the bottom of the Atlantic Ocean. Such underwater cables are very expensive to lay and command a large investment, so Bell Laboratories started looking into the idea of using trans-Atlantic radio telephone calls as a cheaper alternative. But every time they tried making radio telephone calls, they encountered interference. The origin of this interference was unknown; some theorized it could be a competitor deliberately trying to cause problems.

Between the years of 1930 and 1932, a young telephone engineer named Karl Jansky (1905–1950), working for Bell Laboratories in America, was tasked with investigating the interference that was plaguing these long-distance high frequency communications at 20.5 MHz. Jansky's job was to find the source of the interference and come up with a solution to the problem.

Jansky built himself a large antenna that could be moved on a circular track. This contraption was affectionately nicknamed the "merry-go-round" antenna (Fig. 1.2).

Jansky's antenna used two pairs of Ford Model T wheels to support the frame, allowing him to move it single-handedly in the hope that through altering its direction he could narrow down the direction of the interference problem. He found after running his equipment for a short time that the interference seemed to come at regular intervals: every 23 h 56 min and 4 s. This meant it happened nearly 4 min earlier each day.

Fig. 1.2 The "merry-go-round" antenna. Its creator, Karl Jansky can be seen right of centre. (Image courtesy of NRAO/AUI)

This must have left Jansky scratching his head, as Jansky was an engineer, not an astronomer. Most astronomers would have soon realized what this could be. The Earth doesn't rotate on its axis in exactly 24 h, but takes 23 h 56 min and 4 s, and therefore, a point in space such as a star rises approximately 4 min earlier each day. Finally, Jansky concluded that the source of the unknown interference must be coming from the sky and space itself. This was the first such signal received from space.

Thereafter, Jansky tried to narrow down which part of the sky the interference was coming from. This sounds easier than it actually is because, as we will find out later, radio telescopes have very poor angular resolving power. And it is very difficult to pinpoint an exact point in the sky. In order to identify the location of this radio emission as accurately as he could with his equipment, he estimated the beam width of his antenna, then consulted star maps to know what part of the sky—or more importantly which constellation—was passing through the antennas beam at the exact time the interference was detected.

Jansky did this over a number of days, watching to see when the signal received by the antenna was at its strongest. He then came to the conclusion that the source of the radio emissions or interference was coming from the part of the sky that contained the constellation of Sagittarius. The design of the antenna made it impossible to narrow it down any further than this.

Bell Laboratories received Jansky's report and the accompanying realization that the source of the interference was of an extraterrestrial origin, and therefore, nothing could be done to solve the problem. Jansky asked for extra funding in order to investigate this radio source further and try and pinpoint the exact part of the sky from where it was being emitted. The funding to build a larger antenna and more sensitive equipment was refused. Despite his protests that this first-ever extraterrestrial signal warranted close investigation, the funding was never granted.

Jansky published his findings in 1933—a year that would go down in history as the start of the science of radio astronomy.

Jansky died in 1950 from a heart condition at the young age of 44. In his honor, the unit used in radio astronomy to represent the energy or "flux" coming from a radio source is called the "Jansky." A crater on the far side of the Moon is named in his honor, in recognition of his discovery of the first radio source beyond the Earth.

There is a podcast on Jansky available through www.365daysofastronomy. org. This podcast was written and recorded by Dr. Christopher Crockett of the United States Naval Observatory, as part of Dr. Crockett's astronomy word of the week series. The podcast, dated April 11, 2012, is well thought out and presented and worth tracking down. Here is the link: www.cosmo-quest.org/x/365daysofastronomy/2012/04/11/april-11th-astronomy-word-of-the-week-jansky/.

The First Radio Telescope

Nothing was done to follow Jansky's discovery until Grote Reber (1911–2002) came along. As a young electrical engineer, Grote Reber was intrigued by Jansky's results about the mysterious source of extraterrestrial radio emissions and wished to follow up on such research, taking them to what Reber described as a "logical conclusion."

Reber applied several times to Bell Laboratories, but his applications were repeatedly refused, not because of his qualifications or his intention to follow up on Jansky's work, but because they came in the mid-1930s, amidst the Great Depression. Taking matters into his own hands, Reber decided in 1933 to build his own radio telescope (Fig. 1.3).

The antenna was the first radio telescope purposely built to study the sky. The diameter of the dish was 29.5 ft (9 m). Reber constructed it by himself, in his own garden! He built the telescope frame and shaped the dish to give

Fig. 1.3 Grote Reber's homemade radio telescope. (Image courtesy of NRAO/AUI)

it its parabolic shape. He also built all his own receivers. This was long before the miniaturization of electronics with the use of transistors and microprocessors.

The receivers would have been made up of a collection of vacuum tubes (valves) used in the amplifiers and receiver circuits. Vacuum tubes are very fragile, being made of glass, and are difficult to manufacture, so they would have been expensive. Vacuum tubes demand more electrical power to operate and take time to "warm up" before they will operate properly.

When he had finished, Reber had a radio telescope that could only be moved in altitude (up or down), while the azimuth (east-west) movement of the telescope depended on the rotation of the Earth. Thereafter, he started to make sweeps of the sky, but he found that the interference from the local neighborhood, especially from the sparks of vehicle ignition systems, made this very difficult. He tried at different times of the day and found the best time to observe was from midnight to 6 am when most people were in bed. Between these times things quietened down enough to allow him to make useful observations with his equipment. This is still true today.

Reber's day would start in the morning with breakfast and then a 30-mile (50 km) drive to his place of work, in a radio factory. He would work a full day, then drive the 30 miles (50 km) back home. On arriving home around 6 pm he would have his evening meal, then try and grab a few hours of sleep. Just before midnight, he would rise and start observing with his radio telescope from midnight to 6 am. Then, once again, he would eat breakfast and drive to work. And so it went on night after night. He did this for a number of years until he had created the first radio map of all the sky that he could observe from his location. That's dedication!

Reber's map by today's standards would be considered rather crude, but he did confirm Jansky's findings of a radio source in the constellation of Sagittarius. He also found strong radio emissions coming from other areas of the sky and published his findings in 1938.

The American National Radio Astronomy Observatory (NRAO) research center based in West Virginia went on to employ Reber as a consultant from the early 1950s onward. Reber donated his homemade radio telescope to the NRAO, and it is now sited at the NRAO's Greenbank Science Center where it is preserved and maintained. It has been fitted with a turntable that allows it to be turned in azimuth.

In 1954 Reber moved to Tasmania and worked at the University of Tasmania, where he carried on with the study of radio astronomy. He died shortly before his 91st birthday.

There is a very good podcast available on online, which includes an interview with Grote Reber. The podcast can be found at the NRAO's Mountain Radio website. Here is the direct link: https://www.gb.nrao.edu/epo/mra/

Grote.mp3. In the podcast, Reber is interviewed by a member of the Greenbank Radio Telescope staff. Reber talks about how he built his radio telescope and the problems that he had to overcome in doing so.

Here is another link to a podcast, dated August 31, 2009, about Grote Reber: www.cosmoquest.org/x/365daysofastronomy/2009/08/31/august-31st-grote-reber-the-first-radio-astronomer/.

The Outbreak of World War II

In 1939, at the outbreak of World War Two in Europe, a veil of secrecy descended over most of the world.

Countries taking part in the war could ill afford luxuries such as funding research into subjects like radio astronomy. A great deal of academic research was put on hold, but this did not mean that scientists escaped the war—rather, they were reassigned, given new war-related duties such as code breaking or weapons and technology research.

During this period, the United Kingdom invented a new system for "RAdio Direction And Ranging"—RADAR. The first attempts were very basic, tall wooden masts with wires strung between them. These transmitted a radio signal, while a separate mast and antenna were used to receive the returning echo. This returning echo was then shown on an oscilloscope screen, with the operator having to make their best guess regarding what the signal meant and how far away an object was.

As the war raged on, the call for better radar equipment increased and a number of instillations were built to give radar cover to the entire east coast of the United Kingdom. The remains of these installations are still visible along the Norfolk and Suffolk coastlines of the UK.

All the radar installations were in contact with each other by telephone. The advantage of having a larger number of installations was that each station could compare their signals to the ones on either side to see who had the strongest echo. This information was used to give a better and more accurate way of judging an object's distance and direction of travel. It was the job of the station with the strongest echo to call the results in to the senior person so a decision could be made on the best course of action. Despite being very basic, the system allowed enough warning to act against enemy units over the North Sea.

Some radar operators reported receiving a large echo on their equipment that seemed to be coming from over mainland Europe. Was this an attempt by the enemy to block or jam the radar? Had they developed their own radar? If so, what could be done about it?

In 1942 a group of technicians lead by J.S. Hey (1909–2000) was sent to find the origin of this large echo. After an exhaustive search and a thorough examination of the equipment, Hey and his colleagues thought the only possibility left was the Sun. Could the Sun indeed be the source of the signal that the radar operators had reported receiving?

After consulting with astronomers observing the Sun, they confirmed that a large group of new sunspots had appeared on the Sun's disc. This revealed the connection between solar activity and radio emissions from the Sun. Hey didn't publish his findings until 1946, after the end of the war. During the war, Grote Reber had also detected emissions from the Sun, and he published his results in 1944.

Advances in wartime technology also led to the invention of a new piece of equipment, called the cavity magnetron. This was a key development that gave rise to a more accurate radar unit. Using smaller wavelengths that readily reflected off an object, the cavity magnetron could provide greater detail and allow smaller objects to be found and tracked more effectively. This also meant that radar units could be made smaller. Small enough now to be fitted into an airplane, these units gave rise to airborne radar.

It must be noted that this progress came at a price. In the rush to develop the cavity magnetron, safety wasn't given as high a priority, and several people were irradiated, some quite badly. In the workshops where the cavity magnetron was developed, every time it was switched on, people reported that the cheese in their sandwiches started melting. When the radar units were fitted into airplanes, there was very little in the form of shielding, as weight was an issue. The less shielding used, the more fuel and bombs could be carried. The lack of shielding from these new radar units gave rise to similar reports by the flight crew that bars of chocolate they had been carrying with them had melted, providing further evidence that the shielding was inadequate. The workings of the modern microwave oven are a spin-off of the technology used to develop the cavity magnetron, but the shielding problem surrounding the leaking of microwave energy has been adequately addressed.

At this point in the war, radar equipment could be made smaller and more efficiently. Portable anti-aircraft units were now available, and mobile radars mounted on trailers could be towed anywhere they were needed.

While the radar operators were using this new portable equipment, they reported receiving short-duration echoes that lasted from less than a second up to around 5 s. This was another mystery to be solved.

By this time, V1 and V2 rockets were being sent over from mainland Europe. The V2 rockets traveled at high velocities approaching supersonic speeds. (The technology from these V2 rockets would eventually put astronauts on the Moon.) Did these short-duration echoes have anything to do with these new weapons? Hey and his colleagues looked into all possibilities that might explain their existence. His team came up with a different theory—that these echoes could be coming from meteors as they entered the Earth's atmosphere.

After the end of World War II, there were lots of ex-army portable radar units being sold off cheaply to try and recoup some of the money that the war had cost. Hey got hold of some of these units and carried out experiments using the radar units to prove this theory. In 1944–1945, using astronomical data from past meteor shower observations, he proved that the short-duration echoes really did come from meteors. This will be covered later in Chap. 16.

This leads to an interesting story about the Parkes Radio Telescope in Australia. For years, the Parkes Radio Telescope kept picking up short-duration pulses, only fractions of a second in length. They were quite random in nature, with more happening during the day. After an exhaustive search, the source of these pulses still could not be found, until 1 day a research student found the culprit: the microwave oven in the visitor's center! If the user pulled open the door before the microwave had stopped, the minute leak of microwave radiation caused a pulse to show up on the large radio telescope.

There is a good film showing how the first radar was developed. It is available online from movie streaming services, called "Castles in the Sky." Produced by the BBC and starring Eddie Izzard as Robert Watson-Watt.

Project Diana: The Moon Bounce Experiment

Project Diana was an attempt by the United States to bounce a radar signal off the Moon using a modified Second World War radar transmitter and receiver. The name "Diana" was chosen for this project because the goddess Diana in Roman mythology was said to hunt animals at night by the light from the Moon.

The radar equipment used was a SCR-271 early warning radar unit and was army surplus left over after World War Two. A transmitter and receiver were built for the purpose on a site in New Jersey and were modified by Major E.W. Armstrong, an army consultant during the war and famous for pioneering the use of frequency modulation.

The transmitter was designed to transmit at a frequency of 111.5 MHz in 0.25-s pulses. It was hoped that the power of the transmitter, some 3000 W, would be enough to send the pulses through the ionosphere, all the way to the Moon and back to the Earth.

The antenna could only be moved in azimuth, which meant scientists had to wait for the Moon to be in the right part of the sky and within the beam of the transmitter before any attempt could be made. The first successful detection of an echo from the Moon was done by J.H. Dewitt and E. King Strodola in January 1946. The received returning echoes arrived approximately 2.5 s after transmission. This will be covered later in Chap. 17.

The Early Jodrell Bank

In 1945, British scientist Bernard Lovell (1913–2012) had returned to his post at Manchester University in the United Kingdom after his wartime service working on the development of radar. There, he wanted to carry on with research into cosmic rays. Cosmic rays are highly energetic particles that enter the Earth's atmosphere and nowadays are thought to originate from some of the most violent explosions in the universe, such as those from supernovae. Lovell had a theory that went against Hey's theory. He believed that the "meteor echoes" received by the portable radar units were in fact cosmic rays entering the Earth's atmosphere, and radar could be a way to prove this.

Radar and radio observations taken from the city of Manchester were useless because of the level of interference from the inhabitants of the city. A more remote and radio-quiet observing site was needed for Lovell to make his observations. Luckily for Lovell, Manchester University rented a plot of land approximately 20 miles (32 km) south of the city. This area, named Jodrell Bank, was used by the University as a botanical garden.

A team was sent to Jodrell Bank with a portable radar unit. At that time, the site was nothing more than a large, open expanse of land with a small wooden building housing the agricultural workers' tools. The radar unit was fitted with a Yagi antenna and mounted onto a trailer, then set up on the side of the wooden building. This building would become the headquarters for the equipment and staff while at the Jodrell Bank site.

By December of 1945, the team at Jodrell Bank had proved that Hey was right and that the radar echoes did come from ionized meteor trails, and not as Lovell had first thought from cosmic rays. Lovell realized that he would need a more sensitive radio telescope if he was going to detect cosmic rays.

In 1947, the researchers at the newly named Jodrell Bank Experimental Station built a 218-ft (66-m) parabolic reflector made of wire mesh, which pointed towards the sky. There was a mast in the centre of the wire mesh reflector. This mast could be moved around by tightening or slackening a number of guide ropes used to support the central mast, meaning the radio telescope was slightly steerable a few degrees or so in either direction from the zenith. As the Earth rotated, the team built up, strip by strip, a map of the sky that could be seen from their location.

Lovell never detected any echoes from cosmic rays, but the telescope made the first detection of radio emissions from the Andromeda Galaxy. This proved the existence of radio emissions outside our own Milky Way Galaxy.

Pleased by the results from this telescope, but frustrated by the limited steering and the inability of the telescope to cover a greater area of the sky, Lovell put forward plans to design and build a fully steerable radio telescope with a larger collecting area or dish. The new radio telescope design was to have a parabolic dish with a diameter of 250 ft (76 m) made of wire mesh. The structure was to be mounted on its own track so that the whole assembly could be turned in azimuth. Either side of the receiving dish was to be a tower. Each of these towers would have an electric motor and a huge bearing inside a room at the top. These bearings would support the entire weight of the receiving dish assembly and allow the dish to be pointed at any angle in altitude, giving the radio telescope a huge altazimuth mount.

The First Attempt at Interferometry

The construction of a radio telescope is very different from that of an optical telescope. The mirror of a large telescope has to be made to very high specifications and must have the correct shape to bring all the wavelengths of visible light to the same focusing point.

A radio telescope's collecting area can be made of solid sheet metal bent to form a parabolic dish. In some cases this metal can be in the form of a wire mesh. The shape of the collecting area is still important to bring the radio waves to a common focus. Just like an optical telescope the larger the collecting area the better the resolution. Radio telescopes have very poor resolution—as a 100,000 times worse—compared with their optical counterparts. The poor resolution is largely attributed to the fact that radio telescopes work at longer wavelengths, somewhere in the region of 100,000 times longer than those used by optical telescopes.

To put this in perspective: To get the same resolution with a radio telescope as an optical telescope with a 78.7-in. (2-m) mirror, the radio telescope would require a parabolic dish measuring 124.3 miles (200 km) in diameter.

It would be a technological nightmare to try and build a radio telescope with a parabolic dish measuring this diameter. It is more feasible to build a number of smaller dishes that can be placed at intervals, adding up to a collecting area of the 124.3 miles (200 km). If the signals from each individual telescope dish are combined, this gives the same gain as one large dish. This combined array of radio antennae is known as a *radio interferometer*.

If two or more radio telescopes are used at x distance apart, this gives the same effect as having one larger telescope of x diameter. If the distance between the smaller radio telescopes, known as the *baseline*, is increased, then their combined resolving power is also greatly increased. Additionally, the overall resolution of a radio interferometer depends on the frequency at which the system is operating.

Look at the very long baseline array (VLBA). This is made up of ten radio telescopes with dishes measuring 82 ft (25 m) in diameter, spread across the continental United States and Hawaii. This gives the VLBA a baseline of many hundreds of kilometers and a combined resolving power approaching 0.001 arc seconds. At this resolution the VLBA, outperforms the resolution of the largest ground-based optical telescopes.

The effects of the atmosphere greatly limit ground-based optical telescopes to about 0.3 arc seconds. However, modern adaptive optics systems have made great strides in recent years. By flexing the mirrors of the telescope, they can largely cancel out the effects of the Earth's atmosphere.

Interferometry is a reliable observing method, but there is a downside, as all the individual dishes must be coupled together when the signal is received and recorded. This recording must accordingly be time coded. When each recording is played back, the time codes on the recording allow the operator to know the correct part of the recording to match up with recordings made at other telescopes. The timing must be accurate at least to the nearest millisecond or better. Most radio telescopes thusly require an atomic clock, which is accurate to 1 s every 25 million years.

Here is a link to an interesting podcast about Interferometry: http://www. astronomycast.com/2009/03/ep-129-interferometry/.

The World's First Female Radio Astronomers

There is some controversy as to who was the first female radio astronomer, whether it was Elizabeth Alexander (1908–1958) or Ruby Payne-Scott (1912–1981).

Elizabeth Alexander was the first woman to correctly identify radio emissions from the Sun while trying to solve a problem with radar equipment. On the other hand, Ruby Payne-Scott purposely went looking for radio emissions from the Sun and identified two different types of radio emissions. Both scientists deserve mention in this history.

Elizabeth Alexander

Dr. Elizabeth Alexander was born on the 13th December 1908 in Surrey in the United Kingdom. She spent her early days in India where her father was a professor of chemistry. After the First World War, she returned to the UK and finished her secondary education. She was a very bright student and went on to study geology at Cambridge. In 1931 she graduated with first-class honors and 3 years later earned herself a PhD.

Alexander married a New Zealand physicist in 1935, after which the pair moved to Singapore, settled down, and had three children. At the start of 1942, trying to avoid the worst of the war and fearing invasion from Japan, Elizabeth and her children moved to New Zealand. Soon thereafter, she was told the terrible news that her husband had died. From that point forward, New Zealand became their permanent home.

With her academic background, Alexander was appointed to head up a new research department specializing in radar, the Operational Research Section of the Radio Development Laboratory in New Zealand.

In early 1945, a number of mysterious signals were picked up at 200 MHz by operators at the Norfolk Island radar station. Operators noticed that radio interference increased at sunrise and sunset. These mysterious signals became known as the "Norfolk Island Effect," and Alexander set out to uncover what these mysterious signals were.

After a number exhaustive tests, which involved turning the antenna to find the source of the signals and examining recordings, she found that these signals were coming from the Sun—more precisely, a group of sunspots on the surface of the Sun. A year later in 1946, she wrote a paper about her findings that was published in the journal *Radio and Electronics*.

After the war, Alexander found out that she had been misinformed, and that her husband hadn't died. He joined her and the children in New Zealand, whereafter the couple went on to have careers in teaching. In 1958 Alexander was made head of her department, but unfortunately died on the 15th October 1958 after suffering a stroke, just 3 weeks after taking up the position.

Ruby Payne-Scott

During World War II, Australia threw its efforts into erecting a number of radar stations. After the war, towards the end of 1946, physicist and radar specialist Ruby Payne-Scott pointed such wartime radar receivers at the Sun and picked up a number of radio signals. Using interferometry methods,

Payne-Scott was able to improve signal resolution. This led her to identifying two particular solar radio emissions. These emissions are now classified as type I and type III. *Type I* radio emissions appear simultaneously across all frequencies, whereas *type III* radio emissions first appear at higher frequencies and drift to lower frequencies over several seconds.

Within 4 years, Payne-Scott had established herself as a world-class solar astronomer. Later on, in the early 1950s, her career was put in jeopardy for reasons unrelated to her research. At the time, Australian law stated that married women were not allowed to hold a permanent public service position, and there were no laws in place to provide proper maternity leave. For years, Payne-Scott hid her marriage from her employers to avoid these limitations. Later, she was found out, and she lost her permanent position, though still continued to work and receive a salary. This came to an end shortly thereafter in 1951, when she gave up her job several months before the birth of her son.

During the 1960s, Payne-Scott returned to teaching. Her students and colleagues described her as being forthright but an excellent teacher with very high standards. Unfortunately, the onset of Alzheimer's disease put a halt to her career, and she passed away in 1981.

Here is an interesting podcast about Ruby Payne-Scott, dated September 22, 2009: https://cosmoquest.org/x/365daysofastronomy/2009/09/22/september-22nd-the-first-female-radio-astronomer/.

Surveys of the Radio Sky

From the mid-1940s to the late 1950s, several surveys of the sky were carried out by radio telescopes around the world using a number of different frequencies. These surveys identified several areas of the sky from which radio emissions had been received and which required further study. The keyword "area" here is important, as the larger wavelengths used in radio astronomy made it very difficult to pinpoint the exact spot in space the radio emissions came from. These new areas of interest were called "radio stars." The term "radio star" is very rarely used in astronomy today and has been largely forgotten, although it appears in older astronomy books.

With radio astronomy being a new science at the time, no one had any real idea what was producing these radio emissions. Until the accidental discovery of the galactic centre by Jansky in 1933, some thought that radio waves could only be produced by manufactured transmitters.

The hunt was on to find an accurate explanation of what was producing these radio emissions. Many maps were drawn showing the relative position of these mysterious radio stars. Drawings were also made to show the distribution of these radio stars within the galaxy in the hope that this would in some way lead to an explanation.

This quest continued until 1948 at Cambridge in the United Kingdom. At that time, two radio astronomers, Martin Ryle (1918–1984) and Francis Graham-Smith (1923–), were carrying out research into Cassiopeia A, Taurus A and Cygnus A. These had already been reported by J.S. Hey and J. Bolton.

Ryle and Smith were given the task of finding an accurate position for the three radio sources. The pair were using two 27-ft (8.2-m) parabolic dishes from an old German radar, paired together to form an interferometer.

The area around the Crab Nebula was found to be the source of Taurus A. This was already known to be a supernova remnant, but no reason was found as yet regarding how the radio emissions were being produced.

When an accurate position was found for Cassiopeia A, optical astronomers were given the coordinates and asked to observe this area. Optical astronomers reported seeing a faint nebula. Astronomers Walter Baade (1893–1960) and Rudolph Minkowski (1895–1976) at the Mount Palomar Observatory in California were then asked to observe the area and double check the coordinates using the 200-in. (5-m) optical telescope.

A photographic plate was taken of the area using the same coordinates, proving the existence of the nebula. Although the nebula was faint in appearance, the two astronomers at Mount Palomar were able to extract a spectrum of the nebula. Through calculations of the redshifted nebula's spectrum, the nebula was found to be at a distance of 10,000 light years. At this distance it could be placed in an adjacent spiral arm to our own Orion arm, still within the Milky Way Galaxy.

The third source, Cygnus A, was found by Reber and appeared on his radio map of the sky in 1944. Hey knew this area had a nebula that measured several degrees across (NGC 6992, the Veil Nebula) and was thought to be the result of a supernova. The resulting nebula from the supernova covers a large area of sky that is dense with stars from the Milky Way passing through this part of the sky. When Cygnus A was finally tracked down and photographed, the resulting image showed the nuclei of two colliding galaxies, and it became clear that the radio emissions were coming from beyond the visible Veil Nebula.

When a spectrum was taken and the redshift calculated, the two colliding galaxies were found to be at a distance of 550 million light years. This made Cygnus A the furthest extragalactic radio source identified outside our own galaxy.

The 21-cm Hydrogen Line

In 1951, Harold Irving Ewen (1922–2015) and Edward Mills Purcell (1912–1997) discovered the 21-cm line of hydrogen. The scientists used a horn-shaped antenna approximately 72 in. (2 m), square at the open end and reducing in size along its length like a large, square ice-cream cone.

The 21-cm hydrogen line represents a change of energy state in neutral hydrogen atoms. This happens at a precise frequency of 1420.40575177 MHz, which is equal to a wavelength of a little over 21.11 cm.

The physical change to the hydrogen atom is the "spin" of the orbiting electron. This spin quality is the property of electrons and other fundamental particles and describes how they rotate about their axis. The term falls under the umbrella of quantum mechanics. Simply put, a hydrogen atom has one proton and one electron. The spin of the proton and electron can either be the same or oppose each other. If the spin of the electron opposes the spin of the proton, the total energy within the atom is slightly less than it would be if the two had the same spin.

The moment the electron changes its spin, the atom loses energy. This energy has to go somewhere, as per the law of energy conservation, so a low energy photon is released. The energy level of this photon is equivalent to the wavelength 21.11 cm.

This discovery allowed the possibility of studying the structure within the Milky Way, and in particular its spiral shape, because this wavelength can pass through clouds of interstellar dust that visible wavelengths cannot. Astronomers could now observe what was happening inside gas and dust clouds. It thusly had a profound impact on the scientific community, and more locally, forced a change in the design of the radio telescope proposed at Jodrell Bank.

This change in design involved abandoning the original wire mesh and replacing it with a solid dish. The smoother surface of the new solid dish would allow the telescope to be more efficient and to work at a greater range of frequencies including the 21-cm hydrogen line. Before construction of the Mark 1 radio telescope at Jodrell Bank began, the staff had to tackle the problem of where to site the radio telescope, because of the sheer size and weight of the finished structure. It was estimated that the finished telescope would weigh in the region of 1700 tons. Additionally, there were the inevitable financial problems, especially as now the design had changed to a solid metal dish, which added considerably to the total cost.

The project nearly came to a standstill on several occasions, and a plea for money to finish the Mark 1 radio telescope was issued to local newspapers. There is a story of a young school child who, upon hearing about the plight of the unfinished telescope, wrote a letter to Bernard Lovell in which the child expressed concern over the fate of the telescope and the belief that the telescope should be built to explore the universe. The child included spending money within the letter as a contribution in the hope that the telescope would soon be finished. This would surely have tugged at the heart strings of Bernard Lovell, and perhaps it made him redouble his efforts to complete the telescope.

In the end, the funding was received, and by 1957 the Mark 1 radio telescope was finished. At this time it was the world's largest fully steerable radio telescope. In 1987 the telescope was renamed the Lovell Telescope in honor of the now-Sir Bernard Lovell (Fig. 1.4).

By the middle of the 1950s, radio astronomy had taken off, as scientists all around the world had started building all manner of weird and wonderful radio telescopes to explore the universe in radio wavelengths. In the 1950s and 1960s, there were several large radio telescopes at Cambridge and Jodrell Bank in the United Kingdom, Westerbork in The Netherlands, Parkes in Australia, Greenbank in the United States and Arecibo in Puerto Rico, to name a few.

Fig. 1.4 The Lovell radio telescope (formally the Mark 1) at Jodrell Bank Cheshire in the United Kingdom. This image was taken shortly after the surface of the dish had been replaced with a more accurately shaped and smoother surface, which allowed the telescope to be used at even higher frequencies

The Radio Planet

The next discovery in radio astronomy took scientists completely by surprise.

In 1955, Bernard F. Burke (1928–2018) and Kenneth Franklin (1923–2007) of the Carnegie Institute accidentally discovered radio emissions coming from the planet Jupiter. They were testing out a new antenna and working at a frequency of 22 MHz. After using the antenna for the first time, they faced the usual problem of teasing out the signals they wanted from unwanted interference. They worked their way through each of the signals they had received, identifying each one, but there still remained a frustrating bit of "scruff" that couldn't be identified. At first they thought this bit of scruff could be coming from vehicle ignition systems. After an excellent piece of detective work, it was found that this bit of scruff that was in fact the planet Jupiter. This will be covered in more detail in Chap. 12.

Planetary RADAR Astronomy

The planet Venus, named after the Roman goddess of beauty, is the third brightest object in the sky after the Sun and Moon.

The planet is shrouded in a thick blanket of cloud. No views of the surface can be seen—only small amounts of detail in the cloud tops can be observed from Earth-based optical telescopes. Using ultraviolet wavelengths can help reveal detail within the Venusian cloud tops, but this comes with its own problems, as some ultraviolet wavelengths are blocked by the Earth's atmosphere.

It is not unreasonable to assume, because Venus is closer to the Sun than the Earth, that the average temperature would be higher than that of the Earth. Many estimates were put forward regarding the temperature at Venus's surface. In 1952 Harold Urey (1893–1981) estimated that the surface temperature of Venus would be around 127.4 °F (53 °C), but 1951 spectroscope analysis of the atmosphere showed there was a high level of carbon dioxide present. At this time, the effects of such a high level of carbon dioxide in the atmosphere were not fully understood. The phrase "greenhouse effect" hadn't even been coined. One British astronomer, on hearing about the high levels of carbon dioxide, suggested that the oceans of Venus could be fizzy like carbonated water with dissolved carbon dioxide in them.

In 1956, C. Mayer, working with a group of astronomers at the Naval Research Laboratory in the United States, had been studying microwave radar echoes received back from Venus. These radar echoes suggested the temperature could be closer to 626 °F (330 °C). It was the Russians who found the actual temperature of Venus to be 878 °F (470 °C) when they started sending the series of Venera space probes to the planet in the early 1960s.

What they didn't know was the crushing atmospheric pressure on Venus. This pressure is 90 times greater than the atmospheric pressure on the Earth. This would prove to be a serious problem, as several of the earlier Venera probes were crushed by the pressure of the atmosphere before reaching the surface of Venus.

The First Artificial Satellite

Just months after the Mark 1 radio telescope at Jodrell Bank was finished, the Russians planned to launch a series of satellites into Earth orbit. These would be the first artificial satellites to orbit the Earth and were known as the Sputnik series of satellites. (Sputnik means "fellow traveler" in Russian.) These would be a good test of the Mark 1 radio telescope's accuracy and steering capability, as it would monitor the satellites in orbit.

On October the 4th 1957, the Russians launched Sputnik 1. This was to be the start of the Space Race, and the finishing line would be an astronaut or cosmonaut on the Moon.

Sputnik 1 was spherical in shape with a diameter of approximately 23 in. (580 mm). It weighed approximately 185 lb (85 kg). As it flew over the Earth, its distances varied from 150 mile (240 km) to 300 miles (480 km) according to its elliptical orbit. It carried on board a simple transmitter that allowed scientists to track it in its orbit around the Earth. It transmitted on two frequencies, 20 MHz and 40 MHz.

These particular frequencies were thought deliberately chosen to serve two purposes. The first was to study which frequency passed through the ionosphere best—whether that be the higher 40 MHz or the lower 20 MHz. The second reason was that at the time of Sputnik 1's launch, household "wireless" devices could be retuned to pick up the "beep beep" signal from Sputnik 1 as it orbited overhead, thus proving to the world that the satellite was really in orbit. This had the effect of producing panic across the world, as tensions were high among nations at the time, and Russia's technological superiority was seen as an immediate danger.

Laika, the First Living Creature in Space

On November 3rd 1957, Sputnik 2 was launched. This time, a small 3-year-old female dog named Laika was on board.

The name "Laika" means "barker" in Russian. Laika was found roaming the streets as a stray. She was selected from two other dogs and had to endure a punishing training schedule. Before launch, Laika was fitted with a harness that restricted her movements within the capsule. She also had sensors fitted to monitor her heart rate and other vital signs. On entering orbit, the radio receiver on the Earth was switched on and barks could be heard coming from the capsule. Laika had survived the journey into orbit, and at this point everything seemed to be going well.

This was a short-lived victory, as Laika died within a matter of hours. There were many theories concerning how this happened, with many pointing to heat exhaustion as the cause. As the temperature within the capsule started to rise—possibly due to damaged insulation on the capsule or a failure in her life support system—her vital signs started to show that she was becoming more and more distressed. Then suddenly, they started to drop, until there was no pulse rate being received back on Earth.

Laika's death caused outrage and a great deal of controversy throughout the world. The Russians responded by saying it was always intended that she die, as no plans were made to bring her back to Earth. Unsurprisingly, this response just made things worse.

Laika will always be known as the first living creature in space, and also the first to die in space. In April of 1958 the capsule that contained her remains burned up as it reentered the Earth's atmosphere. Many years later, in the late 1980s into the early 1990s, after the fall of the Soviet Union, the Russians released the true list of events surrounding Laika's journey. It was confirmed as first thought: she died of heat exhaustion.

Now, the Russians had proven that it was possible for a living creature to survive in space (if only for a few hours in the case of Laika), and that communication was possible between the spacecraft and the Earth through the ionosphere. The Space Race was truly on.

A number of podcasts on the topic of animals being sent into space can be found on www.astronomycast.org. Here is a brief list:

- Episode 278: Animals in space
- Episode 445: Animals in space Part 1 Insects and Arachnids
- Episode 446: Animals in space Part 2 Mice and other small animals
- Episode 447: Animals in space Part 3 Dogs, Monkeys and More

The First Robotic Probes

Both the Americans and Russians started sending unmanned robotic probes to the Moon, with the aim of finding a suitable landing site for a manned mission.

The Americans had the Ranger series of probes, while the Russians had the Lunik series. Both countries were hoping for close-up images of the Moon even as the spacecraft crashed headlong into the surface. This was done in part to study the composition of the Moon's surface and address vital questions such as: would the Moon support the mass of future astronauts, or would they just sink into the surface, never to be seen again?

In 1959, the Russians launched the Lunik 3 probe. Its mission was to orbit the Moon and take an image of the far side of the Moon. The Moon is tidally locked to the Earth and always shows the same side to our planet. This meant that nobody had ever seen the far side before.

At this time, the Russians didn't have an antenna capable of communicating with the Lunik 3 probe at the distances that would be involved as the probe traveled to the Moon. So the Russians asked Jodrell Bank if they could use the newly built Mark 1 radio telescope to do the job for them. (It is said that Bernard Lovell was at a cricket match when the call came.)

As the Lunik 3 probe went around the back of the Moon, radio contact was lost and all that could be done was wait and hope until it reappeared from behind the Moon. As Lunik 3 came back around, the Mark 1 radio telescope started to receive a signal. Bernard Lovell recognized the incoming signal as a facsimile data signal—at this time, the only technology available for sending an image via radio communication.

Turning this incoming signal back into an image required a fax machine. At the time, the only establishments to own one were newspaper offices. A telephone call was made to a Manchester newspaper office and a fax machine was quickly brought to Jodrell Bank, where it was connected to the radio telescope's feed.

Everyone held their breath as the image was built up line by line. It wasn't long before they realized that it was an image from the far side of the Moon. The following morning, the image was printed on the front of newspapers for the world to see. The Russians were understandably not too happy about the fact the world had seen the image before them. This very nearly caused an international incident between the United Kingdom and Russia.

The Start of the Search for Extraterrestrial Intelligence (SETI)

In late 1959 to early 1960, radio astronomer Frank Drake made an interesting suggestion: If all radio telescopes around the world were listening to the apparent random static from space, wouldn't it be a good idea to listen for signals from an extraterrestrial civilization?

Drake theorized this extraterrestrial signal could be either a deliberate attempt at contacting civilizations across the vastness of space or an accident caused by a radio telescope picking up "leaked" electromagnetic radiation from an extraterrestrial civilization's communications—perhaps their radio or television signals that had entered space. If so, we could eavesdrop on a cosmic conversation.

This idea led to the birth of the Search for Extraterrestrial Intelligence (SETI). Drake carried out the first of these SETI searches in 1960. At the time, he was at the National Radio Astronomy Observatory (NRAO) in Greenbank, West Virginia, United States. This was called Project Ozma, after the princess of the fictitious Land of Oz from the movie *The Wizard of Oz*.

Drake had limited time using the radio telescope, and he needed to figure out whereabouts in the sky he could point the telescope to give him the best chance of receiving a signal. He theorized that if planets did exist around other stars, then a star roughly the same type and age as our own Sun would be a good place to start. His reasoning was simple enough: our Sun is a long-lived and stable star that has existed long enough for advanced life to have developed on orbiting planets.

The next problem was to find suitable stars at the right distance. Radio waves have been leaking into space from the Earth for about 120 years. These waves travel at the speed of light, thus, the greatest distance they could have traveled is 120 light years. If an extraterrestrial civilization 150 light years away was trying to do the same experiment as Drake—i.e. listen in to the radio emissions coming from a place like Earth—then they wouldn't have the faintest idea that the Earth was there for another 30 years, when they at last received our radio waves. Drake therefore had to pick stars close to the Earth to give him greater odds at receiving a signal.

The final problem was to determine what frequency he was going to tune his receiver to. If Drake chose the wrong frequency, then the extraterrestrial civilization could be beaming out radio signals and Drake would have no idea they were there. So if he was going to transmit a radio signal, at what frequency would he chose and why? He theorized the frequency had to be something that was the same throughout the universe and be well known to an advanced extraterrestrial civilization, so he chose a frequency associated with the most abundant element in the universe—hydrogen and the 21-cm line at the frequency of 1420 MHz.

Drake ended up choosing the stars Tau Ceti in the constellation of Cetus the Whale and Epsilon Eridani in the constellation of Eridanus the River. Both stars are of an age similar to that of the Sun, and both are around 11 light years distant. Having chosen which stars he was going to observe and at what frequency he was going to tune into, all he had to do was listen and wait.

From April to July of 1960, Drake tuned the 85-ft (26-m) NRAO radio telescope to 1420 MHz and for a period of 6 h a day recorded the output from the receiver. He and the other radio astronomers repeatedly played the recordings they had made, looking for any signal that might be extraterrestrial in origin, but nothing was forthcoming.

They tried looking for codes within the recordings, such as mathematical connotations like prime numbers and patterns within pulses that seemed to repeat themselves at regular intervals, but still they found no evidence of a signal.

After Project Ozma had taken the first brave steps in looking for extraterrestrial radio signals from alien worlds it made it a feasible scientific goal.

Later, Drake came up with his famous equation:

$$N = R^* Fp\, Ne\, Fl\, Fi\, Fc\, L$$

N = the number of civilizations within our galaxy that communication with may be possible
R* = the average rate of star formation in our galaxy per year
Fp = the number of stars that may have planets orbiting them
Ne = the number of planets around a star that could support life
Fl = the number of planets above that go on to develop life
Fi = the number of planets above that develop intelligent life
Fc = the number of civilizations that develop the technology to make and receive a signal or sign to acknowledge their existence
L = the length of time that such a civilization produces their signal or sign

This equation was a list of variables that were thought to be important to work out the probability of there being a civilization with the technology to make and receive a radio signal or other sign to make their existence known. There is no right or wrong answer to the equation, as the variables can only be estimated with the best knowledge at the time.

From these humble beginnings of one radio astronomer listening to one frequency, the SETI project now has computers scanning millions of signals looking for anything that looks artificial amongst the background hiss from space.

As with most things, science fact is stranger than science fiction.

Here is a link to an interview with Frank Drake: https://www.space.com/28665-seti-astronomer-frank-drake-interview.html.

The Race to the Moon

As the 1960s dawned, US President John F. Kennedy made his famous speech about sending a man to the Moon and returning him safely back to the Earth before the decade was out.

This presented the newly formed NASA with a huge task. Not only did it have to come up with a way to get the astronauts to the Moon and back, but it also had to find a way to communicate with them round the clock while they were traveling to and from the Moon, and also while they were on the surface of the Moon.

This was a problem because no single terrestrial antenna, no matter how big, could communicate with a spacecraft going to the Moon if the antenna was facing away from the Moon when a transmission was needed. They thus needed a number of antennas situated around the world and spaced at suitable intervals, so that as the Earth rotated and each antenna lost sight of the Moon below the horizon, another antenna could take over the communication with the spacecraft.

In this system there needed to be a small overlap between antennas where possible. It need only be a few minutes, but it would allow one antenna to lock onto the signal from the spacecraft before the other antenna lost it. If done correctly, the changeover could take place without a break in transmission. With the cooperation of a number of countries, several suitable antennas around the world were found or built. This came to be known as the NASA Deep Space Network.

By the start of the Apollo missions, NASA had its Deep Space Network up and running, and round-the-clock communications with spacecraft traveling to and from the Moon were now possible.

Everything with the NASA Deep Space Network was working fine as the three astronauts Neil Armstrong, Buzz Aldrin and Michel Collins approached the Moon and planned to go into orbit around it. At least this was the case, until the night before the famous first Moon walk was due to take place. At the time, there was a problem with the Parkes radio telescope in Australia.

This was made all the worse by the accelerated schedule as Armstrong and Aldrin, now in the lunar module, were flying down to the surface of the Moon. Once Armstrong made the landing on the Moon, they found that it was scheduled for the two astronauts to have a sleep break. Armstrong understandably overruled this sleep break. They had just landed on the Moon, who would feel like sleeping?

The radio telescope at Parkes managed to find the signal just in time to watch Neil Armstrong descend the ladder and take his first step onto the Moon's surface.

These events are dramatized in the 2007 movie *The Dish*. Here are two podcasts discussing the Deep Space Network:

- https://cosmoquest.org/x/365daysofastronomy/2020/01/12/jan-12th-nasa-deep-space-network-turns-50-hubble-finds-three-surprisingly-dry-exoplanets/
- http://www.astronomycast.com/2019/03/ep-521-the-deep-space-network/

The First Quasar Is Discovered

In the early 1960s, a series of experiments with the Mark 1 radio telescope at Jodrell Bank and other small telescopes at increasingly greater distances from the Mark 1 served to further develop and refine the system of radio interferometry.

Between the years 1960 and 1963, using the method of interferometry, a sky survey conducted by Cambridge University found that a number of the most powerful radio sources in the sky had quite small angular sizes.

One particular object of interest was given the name of 3C273. The first 3 denotes that it was found during the third survey; the "C" stands for Cambridge, and 273 means it was the 273rd object found during the survey.

The exact position couldn't be narrowed down enough to make an observation with an optical telescope. Armed with this information, the Parkes radio telescope in Australia stepped in and took on the challenge. It used occultations of the Moon—watching the Moon transit across the sky and taking its exact position when the signal being received by the radio telescope was lost behind the Moon's disc. Since the Moon has an angular measurement of approximately half a degree, this was enough to narrow the search area so that the large optical telescope on Mount Palomar, California was able to see it optically.

This powerful radio source was given the name *quasar*, for "QUAsi-Stellar rAdio souRce." Quasars emit energy over a wide range of wavelengths, including optical. They are thought to inhabit the centers of very remote galaxies. When quasar 3C273 was finally found optically, its spectrum was discovered to be very unusual and unlike anything else ever seen before. It had very wide emission lines within its spectrum that couldn't be identified, until an astronomer by the name of Marten Schmidt realized that the spectra were massively shifted towards the red end of the spectrum. When the redshift was worked out, it was found to be 0.6. This meant it was traveling at an extremely high recessional speed, far faster than anything recorded before, therefore the object must be a great distance from our galaxy.

Quasar 3C273 can be seen optically with amateur sized telescopes. It has a magnitude of +12.9. Its position is RA 12 h 29 m 7 s Dec +2 deg 3′ 9″.

The First Interstellar Molecule Is Discovered

In 1963 the first interstellar molecule was discovered by S. Weinreb and a team of radio astronomers. The molecule was found within the absorption spectrum of Cassiopeia A. It was hydroxyl radical (HO).

The hydroxyl radical molecule is the neutral form of the hydroxyl ion. Hydroxyl radicals are highly reactive and for this reason are usually short-lived. If these molecules were introduced into the Earth's atmosphere they would instantly react with it. But in space, there is very little of anything for the hydroxyl radical molecule to react with, apart from the odd gas molecule, therefore these molecule have a much longer life expectancy.

Like the 21-cm hydrogen line, this new discovery gave radio astronomers another wavelength, 18 cm (1665.4018 MHz), in which they could probe further into the interstellar gas clouds. This was useful for exploring areas such as the Orion Nebula that have active star formation regions, and other areas around the expanding gases and material of supernova remnants.

Additionally, the discovery of the hydroxyl radical molecule proved to astronomers that atoms could come together in space and form molecules. This incited questions about the ability of more complex molecules to form and exist in the extreme cold vacuum of space.

RADAR Used on the Planet Mercury

The planet Mercury has always been a problem to astronomers because of its small physical size of 3032 miles (4880 km) in diameter. Additionally, Mercury has a highly elliptical orbit, which causes a huge change in the angular size of the planet as observed from the Earth's surface. The planet's proximity to the Sun also makes it difficult to observe.

At greatest elongation, Mercury is no more than 28° away from the Sun as seen from the Earth. For this reason optical astronomers have never seen Mercury in a truly dark sky, but instead only at dawn or dusk depending on where Mercury is in its orbit.

It is dangerous to sweep the sky looking for Mercury, as the Sun may accidentally be brought into view through a pair of binoculars or a telescope. This would instantly damage the observer's eyesight, potentially permanently. Damage could also occur to light-sensitive equipment such as the chips used in CCD cameras.

When the planet Mercury is viewed in a twilight sky, it is always low down in the thickest part of the atmosphere where the "seeing" is at its worst. The planet appears to have phases similar to that of the Moon and has a pinkish hue to it. Surface detail is unlikely to be seen with an amateur telescope, and imaging of the planet through one is no better. High-magnification images of the planet show that Mercury has a heavily cratered surface like the Moon.

These problems observing the planet with optical telescopes meant that for a long time, the true rotational period of the planet was not known and could only be estimated. The rotational period of Mercury was first thought to be the same as its orbit of 88 days, in the same way that the rotation of the Moon is roughly the same as its orbit around the Earth. This can be shown as a ratio of 1:1.

In 1965, using the 1000-ft (305-m) Arecibo radio telescope in Puerto Rico, radio astronomers Gordon Pettengill and Rolf Dyce carried out a number of experiments to bounce radar off the planet to find Mercury's true period of rotation. It was hoped the returning echoes would prove or disprove once and for all the theory of the 88-day rotation of Mercury. From this, it was found that the planet had a truly unique period of rotation, unlike any other planet within the Solar System. Mercury has an orbital resonance ratio of 3:2, meaning the planet rotates three times on its axis for every two orbits of the Sun. When this orbital resonance was calculated, it was found that Mercury takes 59 Earth days to rotate once on its axis.

This unique rotation, coupled with its highly elliptical orbit (more eccentric than any other planet in the Solar System), would make for some very interesting effects if it was possible to stand on the surface of Mercury. The Sun would be seen to rise, but as the Sun got higher in the sky it would appear to get smaller in size, due to the planet's highly elliptical orbit. Then the Sun would appear to backtrack across the sky in the same direction from which it rose and set, and then the Sun would rise again, but this time it would travel right across the sky with the same apparent change in size, and set. This would be like seeing a double sunrise and one sunset.

Here is a link to a podcast about RADAR and how it can be used in astronomy: http://www.astronomycast.com/2011/10/ep-233-radar/.

The Cosmic Background Radiation Is Discovered

In the 1960s, there were several theories that attempted to explain the origin of the universe. Some had an element of science to them, while others were rather strange.

The two leading theories regarding how the universe came into being were the steady-state theory and the Big Bang theory.

The *steady-state theory* stated that the universe has always existed and will carry on existing forever without noticeable change. Stars will die but new stars will be born to take their place in an endless cycle.

The *Big Bang theory* states that the universe had a beginning and will have an end. The universe is said to have begun from a vast explosion from which all the matter in the universe came. According to the theory, this explosion was the start of everything, including time and space, and is thought to have happened 13.8 billion years ago. It predicts the end of the universe could come in the form of a big crunch when the expansion of the universe stops and gravity pulls everything back together, or the universe could carry on expanding forever, ending its days as a cold, dark place devoid of heat and light.

Fred Hoyle was a well-respected British astronomer and writer. Hoyle was greatly in favor of the steady state theory. In a series of radio broadcasts, Hoyle shared his thoughts about the universe and referred to the theory that it started with a large explosion as a "big bang" idea. This off-the-cuff comment stuck and was taken up by other astronomers and physicists, such as Prof. Stephen Hawking, as the name to describe their opposing theory.

The battle of the steady-state and Big Bang theories raged on until two scientists entered the scene: Robert Wilson and Arno Penzias, working for

Bell Laboratories in Holmdel, New Jersey. (This was the same Bell Laboratories that Karl Jansky worked for when he found the radio source in the constellation of Sagittarius.)

Wilson and Penzias were using an interesting type of radio telescope, the 20-ft (6-m) horn antenna. Its design was so unusual that it was nicknamed the "sugar scoop" because of its striking resemblance to a huge scoop. Wilson and Penzias were originally going to use the horn antenna as a way of receiving the echoes from radio waves bounced off of balloon satellites. Balloon satellites were sometimes referred to as "satelloons;" these satellites had been inflated with gas after being placed into orbit.

As soon as Wilson and Penzias switched on the receiver, they picked up all the usual sources of interference along with an annoying hissing sound that seemed to always be present. They needed to know whether or not they were receiving the echoes that they wanted from the balloon satellites, so the two scientists had to painstakingly eliminate, or at least account for, every source of interference their antenna and receiver was picking up. So the two scientists set about eliminating each source in turn, but there was a source of interference, a steady hiss, that was always there and wouldn't go away.

Their first thought was that the hiss could be interference from the nearby city, but this was soon discounted, as no matter what direction the antenna was pointing, the hissing noise was always there. What was more puzzling was that no matter where the antenna was pointing, the noise seemed to be at a constant level. This meant that the hissing noise wasn't coming from one particular source, and therefore it had to be something that was the same in every direction. This led Wilson and Penzias to think the hiss could be coming from the receiver itself.

Maybe this hissing noise was from a noisy power supply, or the electronics within the receiver producing thermal noise. The two then tried using liquid helium to cool the receiver to -452 °F (-269 °C), just 4° above absolute zero, to remove all the thermal noise from the receiver circuitry. They hoped this would account for the problem, but the hissing was still there even at the extremely low temperature!

After the cooling of the receiver failed to remove the hissing, they turned their thoughts skyward to eliminate radio emissions from the galaxy, the Sun, even the Earth. Still, the hissing noise was present.

The two then took a less technical approach to remove the hissing noise.

A family of pigeons had decided it would be a good idea to make their nests inside the horn antenna. With birds comes the problem of their droppings, and lots of them. Wilson and Penzias promptly frightened the family of birds away and took a broom to the inside of the horn antenna to clear it from the countless bird droppings. No sooner had the two scientists frightened the birds away than the pigeons came back.

This battle raged on for several weeks, and the hissing noise from the receiver never stopped. It didn't seem to matter whether the pigeons were there or not. Wilson and Penzias discussed with other scientists their problem. A scientist named B.F. Burke (mentioned earlier for picking up radio emissions for Jupiter) told the pair about a paper he had recently read: if an explosion as great has the Big Bang had taken place, there must be a remnant or echo to it, and that this echo may possibly be found within the wavelengths they had been using. The pair realized that they had accidentally found the faint echo of the largest explosion in history, the Big Bang.

This gave the Big Bang theorists all the proof they needed, and the steady-state theory soon fell out of favor, although Fred Hoyle refused to believe it right up to his death. Robert Wilson and Arno Penzias were awarded the Noble Prize for Physics in 1978 for the discovery of the Cosmic Microwave Background (CMB) radiation.

This hissing noise is all that is left of the Big Bang, a faint ghost from the start of the universe. The CMB radiation can easily be picked up on a television set or a radio by tuning them to an unused channel. Approximately 1% of the white noise or static that can be heard is the Cosmic Microwave Background radiation from the big bang.

Here is a link to an interesting podcast about the Cosmic Microwave Background: http://www.astronomycast.com/2006/10/the-big-bang-and-cosmic-microwave-background/.

Little Green Men (LGMs)

In 1965 in a field outside the city of Cambridge, United Kingdom, a radio telescope was built. Two years later in 1967, it would discover one of the strangest objects ever that would push the laws of physics to their breaking point.

At the time, Antony Hewish and a group of research students, one of whom was Jocelyn Bell, wanted to build a radio telescope to carry out radio observations of the sky. As always, funding for such a telescope wasn't very forthcoming. Hewish, working with the budget he had, settled on a multiple dipole array for the antenna. To save money, it was to be built by Hewish and his group of students.

This work consisted of hammering a large number of wooden stakes into the ground and stretching wires between them to form the antenna. Hewish and the research students took turns hammering the wooden stakes into the ground and fastening the wires to them. Then all the wires had to be soldered together to complete the circuit and turn this jumble of wires into an antenna. After 2 years of work, they had built a radio telescope made up of wooden stakes and a jumble of wires that crisscrossed the field. The antenna covered an area of approximately 19,360 yd^2 (16,200 m^2).

After its completion, Jocelyn Bell was chosen to take charge of the running of the receiver and chart recorder. At a later date, in an interview, she recalled how the chart recorder was producing 96 ft (29 m) of tracing per day that she then had to examine. Within months of the radio telescope being switched on, Bell had several miles of chart recording paper. On examination, she noticed that at certain times of day a signal had been recorded on the chart paper, and it repeated every one and one-third seconds as regular as clockwork.

This pattern was so regular, it crossed her mind that it could be an artificial signal, possibly from an extraterrestrial origin. For this reason, Bell gave the signals the nickname LGMs, short for Little Green Men. The next thing was to find out where the LGMs signals were coming from. (As mentioned earlier, radio telescopes are notorious for having poor resolution). This was achieved by taking note of what part of the sky was moving through the antenna beam when the received signal was at its strongest. The width of the antenna beam with this type of design would have been quite large, so only an approximation of where the signal was coming from could be made.

All manner of theories as to what this signal could be were suggested. Hewish theorized the object must be small in physical size or it would have been observed already, and it had to be very powerful to generate such a signal. Hewish made enquires and asked a number of theoretical physicists if any had theories regarding what this very small and powerful object could be. The answer that kept coming up was a neutron star. Up to this point, neutron stars had only existed within the minds of the theoretical physicists, and there was no physical evidence that they existed or that they could exist.

A neutron star is an extremely small and dense object—so dense it is thought that a sugar-cube-sized piece of matter from a neutron star would have a mass of 100 million ton.

Neutron stars are formed when a massive star reaches the end of its life and explodes as a supernova. As this outward explosion takes place, the star blasts off its outer layers. The outward explosion masks an implosion that compresses the leftover core. The compression forces are so high within the core that the empty space that is usually present in all atoms between the orbiting electrons and the atom's nucleus is lost. The electrons and protons are merged into neutrons, hence the name. All the neutrons are forced together so there is almost no room between each neutron. This is what accounts for the star's very high density.

It is thought that a neutron star with the same mass of the Sun would be compressed down to a diameter of around 12 miles (20 km). Stranger still, the surface of a neutron star is thought to be a form of iron, but 10,000 times denser than iron found on Earth. Below this crust of super-dense iron is thought to be a mantel of liquid neutrons. What lies at the core of a neutron star is still under debate, because it is so dense within the core that the laws that govern physics start to break down. Some theorists believe it could be a new exotic form of neutrons, yet to be classified.

Neutron stars rotate very quickly on their axis. A better word to describe this very quick rotation would be *spin*. The spin of a neutron star can be many times per second. This is due to the conservation of angular momentum law, in the same way an ice skater on the ice will spin faster if they bring their arms in closer to their body.

The gravity on a neutron star is immense. If the same gravitational force were applied to the Earth's atmosphere, the top of the atmosphere would be <0.25 in. (6 mm) from the Earth's surface.

Any magnetic field that the star originally had would now be concentrated within this small, fast-spinning corpse of a star. The concentrated magnetic field then produces two high-powered beams of energy coming out of the magnetic poles of the neutron star, in the same way that the beams of a lighthouse sweep out to sea. If one of these beams from the neutron star then sweeps across a radio telescope, it will produce a signal.

It was found that the signal from the radio telescope coincided with a part of the sky that contained the constellation of Taurus. Optical astronomers were then asked if there was anything in that particular part of the sky that warranted further investigation. The answer was the Crab Nebula, a supernova remnant.

Optical astronomers already knew about the Crab Nebula being a supernova remnant, as the cloud of gas that makes up the nebula is still expanding. They even had a date at which the supernova explosion took place. This was the year 1054 AD. This information had been found in an ancient Chinese manuscript from that time, which recorded the event. That year, Chinese astronomers detailed the appearance of a "guest star" that was observed in the same part of the sky where the Crab Nebula is now. They described it as shining for 22 months before fading. This guest star could be seen in daylight like the planet Venus. For a time, it was possible to write by its light at night.

With this new information, optical astronomers observed the Crab Nebula again, but this time with the use of high-speed photography. It was found that the central star was the corpse of a long-dead star, spinning at 30 times per second. This is what Bell had observed on the chart recording from the radio telescope.

Up to 1967, this was the most powerful thing ever photographed in the universe. These strange objects were given the name *pulsars* because of the pulses they produced on a chart recorder. Antony Hewish and Martin Ryle both shared the 1974 Nobel Prize for Physics for the discovery of the first pulsar. Ryle was chosen to share the prize, as in the early 1960s Hewish and Ryle had developed the technique of aperture synthesis, which is a form of interferometry. There was controversy over Hewish sharing the Noble Prize with Ryle for this discovery. It was argued that Jocelyn Bell should have been credited with her part in the discovery. After all, it was Bell who had found the original signal, realized its importance and brought it to the attention of Hewish. Fred Hoyle, the astronomer who favored the steady-state theory of the universe, added his support for Bell to receive proper recognition.

Unfortunately, Bell was not given the credit until much later: in 2018, Dame Jocelyn Bell Burnell was awarded the "Special Breakthrough Prize in Fundamental Physics." The prize came with a cash prize of three million dollar (£2.3 million). Dame Jocelyn decided to donate the prize money to help female students wishing to have a career in physics.

Here is a link to a short video explaining the prize: http://www.youtube.com/watch?v=Y66GgSSnl8c.

There is an interesting podcast available here: https://cosmoquest.org/x/365daysofastronomy/?s=joceryn+bell.

To hear audio clips from pulsars, visit: www.astrosurf.com/luxorion/audiofiles-pulsar.htm.

Once there, click on an icon to hear a short audio clip of a number of different pulsars. To the right of the icon is a brief description of the pulsar and its origin. The audio sound between different pulsars is quite marked. Some sound like the speeded up ticking of a clock, but the millisecond pulsars sound like a hammer drill boring a hole in a wall and are quite ear-piercing.

We're Over Here

The idea of advertising the existence of life on Earth is not a new one and goes back some 60 years before Hertz discovered radio waves in 1880.

In the 1820s, German mathematician Karl Gauss had the idea of cutting down trees in a forest to form large geometric shapes such as squares, triangles and circles, which would be large enough to be seen from space. The theory was that any alien looking at the Earth would see the geometric shapes and realize this must have been done deliberately.

In the 1840s, Viennese astronomer Joseph von Littrow put his own twist on this idea. He proposed using the same geometric shapes, but this time digging an outline of these shapes in the dessert. Each shape was to be about 20 miles (32 km) in diameter. Around the perimeter of the shape, a trench would be dug. These trenches would be filled with kerosene (paraffin) and set on fire.

The planet Mars has always been an object of fascination among astronomers and the public, even before the infamous canals described later in the nineteenth century.

In the 1860s, French physicist Charles Cros had the idea of putting up giant mirrors across France in the shape of the big dipper (the plough) in the hope that sunlight would be reflected back into space from the giant mirrors, and any Martians looking at the Earth would see this shape and recognize it as a star pattern.

Now fast forward a 100 years to the invention of the radio telescope.

The giant Arecibo radio telescope in Puerto Rico is an extremely sensitive radio receiver—so sensitive that it could pick up the signal from a cellphone used on the planet Jupiter. In 1974, the giant dish was going to transmit the most powerful transmission ever made in history, deliberately beamed into space.

This signal was meant to be a great cosmic hello, letting any extraterrestrial civilizations within reception of the transmission know the planet Earth was here. The idea was met with mixed reactions. The main cause for concern was: what if a hostile extraterrestrial civilization received the transmission? The Earth has just screamed out "we're over here!" Some believed that it would be a much better plan to just eavesdrop on any cosmic communications first, to see if potential extraterrestrial civilizations were friendly or not.

It was pointed out to anyone who opposed the transmission that it would take thousands of years to arrive. So, it could be expected to take just as long to receive a reply. It was hoped that if an extraterrestrial civilization had the technology to travel the thousands of light years to reach the Earth, they would be of good character and wouldn't want to take over the planet Earth.

The transmission was due to take place from Arecibo on November 16th 1974. It was part of the celebrations marking a major upgrade to the radio telescope. With the upgrade of the radio telescope, its new more powerful transmitter now measured in megawatts. The 1000-ft (305-m) dish was capable of beaming out a transmission to a relatively small area of the sky. This meant that all the power could be concentrated into an outgoing signal equivalent to many megawatts of power.

The signal was going to be sent towards the globular star cluster known as M13 in the constellation of Hercules. The star cluster M13 is the best globular star cluster in the northern hemisphere. It is visible as a faint, misty patch to the unaided eye, but a telescope shows the full beauty of the cluster. This target was chosen because of the great volume of stars within the cluster. It is thought to contain around 300,000 stars or more and is about 25,000 light years away. With the cluster's large number of stars, it was hoped there would be a better chance of finding a planet orbiting one of these stars.

The target had been chosen, and the most powerful transmitter on the Earth was ready, but what was going to be broadcast?

A message needed to be written that if received and decoded would prove to an extraterrestrial civilization that there was other life in the universe.

Frank Drake of SETI, whom we have encountered before, wrote a draft of what he thought should be in the message. With the help of Carl Sagan (Sagan would be involved at a later date with the gold discs on the two Voyager space probes), and aided by other astronomers and mathematicians, they pooled their knowledge and wrote a special mathematical message.

The message was to be broadcast in the most basic fashion possible: that of binary code ones and zeros, and made up of 1679 bits. This number was chosen because it is the product of two prime numbers, namely 73 and 23. This meant the message could either be written as 73 columns and 23 rows, or 23 columns and 73 rows. If the message was received by an extraterrestrial civilization, it was hoped that by using mathematics the message could be reassembled in one of two ways. If the wrong way was chosen, using 73 columns and 23 rows, the message would look like a random pattern, but if the right way was used, 23 columns and 73 rows, an image would be formed that couldn't be mistaken for anything other than a message from another civilization.

The message contained images of:

1. A human being
2. A radio telescope dish
3. A double helix to represent DNA
4. The numbers 1–10
5. A representation of the Solar System
6. Physical and biological information including a list of chemical elements

These images were basic in appearance and were made up of large square pixels similar to the very first video arcade games back in the 1970s.

The entire message was broadcast in <3 min. It is now traveling at the speed of light towards the star cluster M13 and due to arrive in just under 25,000 years. If any civilization receives the message and understands it, then decides to send a reply, it will take another 25,000 years for the reply reach the Earth.

As everything in space is moving, in some cases very quickly, the star cluster M13 will have moved its position relative to the Earth by the time the transmission gets there and vice versa. The Solar System and the Earth will have moved partway around on its 220-million-year orbit of the galactic centre in the 50,000 years it would take to get a reply.

The WOW! Signal

At the Ohio State University Radio Observatory in the United States was one of the longest-running SETI programs conducted, running for 22 years from 1973 to 1995. A radio telescope had been built in Delaware, Ohio to survey the sky for wideband extraterrestrial radio sources. Construction started in the late 1950s and finished in 1963. The radio telescope was given the nickname "Big Ear"—surely the most appropriate name possible for a radio telescope. The Big Ear telescope was in a fixed position and relied on the rotation of the Earth for it to scan the sky. This meant that it could monitor a point in space for approximately 70 s before the rotation of the Earth moved the object out of the telescopes beam.

The morning of August 15th 1977 would start the same as every morning for radio astronomer Jerry R. Ehman. Little did he know that while using the Big Ear radio telescope, he would receive a signal from space that to this day remains a mystery. This signal became known as the *Wow! signal*.

What made the Wow! signal so special was the bandwidth at which it was received. Natural radio emissions are usually at a wide bandwidth and can be received at a number of different frequencies simultaneously. A manufactured signal, such as one produced by a radio transmitter at a local radio station, is a much more *narrowband signal*. This allows the listener to tune in to a number of different stations clearly and without any overlap in frequency between each station.

Wherever and whatever transmitted the Wow! signal seemed manufactured, as the signal had a very narrow bandwidth. When this signal was received it was found that it was at a much higher level, something approaching 30 times the level normally perceived as the average level for the background noise of space. This was beginning to look more and more like a deliberate transmission.

The next interesting thing was the frequency at which the signal was received. Remember from earlier that when Frank Drake carried out the first SETI search back in 1960s, he chose to listen for extraterrestrial transmissions at the 21-cm hydrogen line at the frequency of 1420 MHz. He chose this because hydrogen is the most abundant substance in the universe.

The Wow! signal had a narrow bandwidth and was picked up at a frequency very close to 1420 MHz of the 21-cm hydrogen line. All these factors seemed to be too much of a coincidence for anything that could have been produced naturally.

The signal was received for approximately 72 s. This coincided exactly with how long the signal was within the beam of the radio telescope before the Earth's rotation moved the telescope and the signal was lost.

On the original printout, Ehman wrote the now famous "Wow!" at the side of the signal. Investigations were carried out later to prove that this signal did in fact originate from deep space and not anything within the Solar System, and had nothing to do with any manmade source or the military. Manmade sources of interference at this frequency are highly unlikely, as this particular frequency is one of a number of protected frequencies that are not used because of the significance to radio astronomy (see Chap. 19 for a list of protected frequencies).

The signal came from within the constellation of Sagittarius. Ehman made several more attempts over the next few weeks and months to try and receive the signal again, but to no avail. Several other radio astronomers have tried over the years to receive the signal using more modern and more sensitive radio telescopes, but every attempt to date has failed.

The year 2017 marked the 40th anniversary of the Wow! signal. Exactly what the signal was and the story of its origin remain a mystery. All that is known with any certainty is that it was extraterrestrial and came from the direction of the constellation of Sagittarius.

There is a tradition amongst SETI observers to always have a bottle of Champagne chilling in a refrigerator at the radio telescope observatory where they are carrying out their observations, just in case they do receive an extraterrestrial signal. Whether this tradition came about because of the WOW! signal is not known, but Jerry Ehman could be forgiven if he opened a bottle that night, considering all the factors involved.

Here is a link to hear the audio of the WOW! signal: https://seti.net/indepth/wow/wow.php.

Other Unknown Signals from Space

The Wow! signal was a well-publicized event. But strange signals from space are more common than people think, and caution must be taken in their investigation. A signal may look like it could be a manufactured one, but nature has a habit of proving otherwise.

The list below gives four examples of mysterious signals received since the Wow! signal:

1. **Lorimer bursts**: These are very short-duration, high-power signals. Scientists going back over old data have found Lorimer bursts that originally were missed. Given that they may have been recorded but not identified, nobody knows when they were first picked up. Their origin still remains a mystery.
2. **Space roar**: This signal was picked up from a receiver on a high-altitude balloon. No one knows from which particular direction it came. It was described as being a loud hissing noise but five to six times louder than the background levels. No one has any idea what this could be.
3. **SHGb02=14a**: This signal was flagged by SETI and has been observed three times for about a minute each time. What makes the signal even more interesting is the frequency it was picked up on, which is very close to the 21-cm hydrogen line proposed by Frank Drake in the 1960s.
4. **Fast Radio Bursts (FRBs)**: These are similar to Lorimer bursts. They are powerful short-duration bursts of energy, but unlike Lorimer bursts, which are only picked up the once, FRBs are repeat offenders, being picked up as many as three times.

Here is a link to a video clip entitled ten unknown signals from space: https://www.youtube.com/watch?v=6mwKJAb211.

Contact

In the event that an extraterrestrial signal is received, SETI has produced a protocol for what should follow, which is summarized as follows:

1. On receiving an extraterrestrial signal, it must be confirmed by another radio telescope installation.
2. Once proven authentic, the heads of state of the country that first received the signal should be told of its existence.

3. Next, the heads of state from that country should inform the United Nations of the signal, so that a decision can be made about how to proceed and whether to reply.

Such a decision will probably have to be taken with the advice of scientists and philosophers, depending on where the signal originated and the distances and timescales involved.

Any potential conversation with an extraterrestrial civilization could theoretically span many generations. So the children of future generations will need to be told of the existence of the signal and a rough timescale in which to listen for a reply.

Modern SETI

From its humble beginnings when Frank Drake and his colleagues were tuning into just one frequency, SETI now searches millions of signals simultaneously.

One drawback is that this produces huge amounts of data. If we include data from radio telescopes around the world, then the amount of data that needs to be processed becomes a huge challenge.

In May 1999, with the explosion of the internet, a project called "Seti@ home" came online. By downloading a small amount of software from the https://setiathome.berkeley.edu/ website, a personal computer can help search for signals from ET.

The whole sky is broken up into thousands of small areas like the pieces of a jigsaw. In this system, each computer is sent a small piece of data from the SETI site. This is real, unprocessed data from radio telescopes around the world. The computer processes this data and then returns it back to SETI. The software works when the computer is not being used and runs quietly in the background, as the data is downloaded and returned automatically.

Here is a link to a video clip explaining the process: https://www.youtube.com/watch?v=alJV5aQR68.

People around the world have signed up to this project and donated processing time from their computers to aid the search. Just a couple of hours a week is all it takes. The computer processing time when this video was made runs over 1000 years per day; this processing power has increased vastly since 1999.

This is a nice little radio astronomy project to take part in. It takes <5 min to sign up. Just download the software, open the file, and install it—it's as simple as that. There is also a smartphone app that can be downloaded from the same website, which will work the same way as the computer software.

The Future of Radio Astronomy

Radio astronomy has been and still remains a useful tool for making observations of objects that could not otherwise be observed within optical wavelengths. It has proven its worth in a very short time by demonstrating the existence of some truly remarkable and exotic objects within our galaxy and beyond.

New, larger, and more powerful radio telescopes have now come into service. In 2016, the 500-m Aperture Spherical radio telescope (FAST) in China came online and is now the largest radio telescope in the world, measuring 1640 ft (500 m) in diameter. This dwarfs the 1000-ft (305-m) Arecibo radio telescope.

Here is a link to an interesting video about the FAST radio telescope: https://www.youtube.com/watch?v=XM1YOBzahz.

The Square Kilometer Array (SKA) is being built in South Africa and Australia. When finished, the array will be the largest and more sensitive radio telescope array in the world.

Here are links to interesting videos about the SKA in…

- Australia https://www.youtube.com/watch?v=cJB5xfPTzsw
- South Africa https://www.youtube.com/watch?v=gp4AZwhZQ

SETI's Allen Telescope Array is a number of radio telescopes grouped together to give the resolution of one large telescope. Its aim is to get around 350 such telescopes working together, still following the plan first put forward by Frank Drake to look at the closest Sun-like stars.

Here is a link to an interesting video about the Allan array: https://www.youtube.com/watch?v=XM1YOLBzahA.

With these advancements in radio astronomy technology, the next big discovery could come any day.

Chapter 2

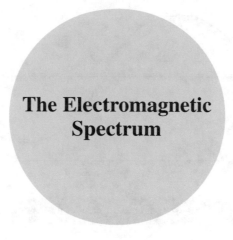

The Electromagnetic Spectrum

Wavelength and Amplitude

Wavelength is a measure of the distance between the wave patterns repeating themselves.

Figure 2.1 shows the image of a sine wave. The graph shows time across the bottom axis, wavelength across the top axis and amplitude on the vertical axis. The graph shows one complete oscillation of the waveform before it starts to repeat itself—this is known as the wavelength of the waveform. Wavelengths can be measured in millionths of a millimeter or thousands of kilometers.

Amplitude is a measure of the signal strength of the wave. The higher the peak-to-peak distance, the more power or amplitude the signal has.

Frequency

Frequency is the measure of how many repeat patterns of the waveform travel past a fixed point per second. For instance, if we use Fig. 2.1 and say the timeline across the bottom is equal to 1 s, then this means the frequency

---

The original version of this chapter was revised. The correction to this chapter is available at https://doi.org/10.1007/978-3-030-54906-0_20

S. Arnold, *Radio and Radar Astronomy Projects for Beginners*, The Patrick Moore Practical Astronomy Series, https://doi.org/10.1007/978-3-030-54906-0_2

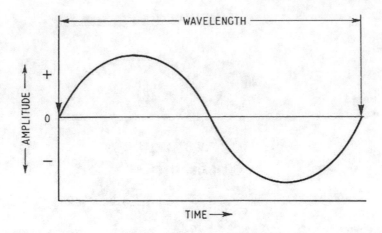

Fig. 2.1 Image showing wavelength and amplitude of a sine wave

of the waveform would be 1 Hertz (abbreviated to 1 Hz). If the pattern of the wave is repeated twice a second, we would say the frequency is 2 Hertz (2 Hz), and so on.

Look at an analogue radio and the frequency marked on the slider for example. A frequency of 100 MHz (Mega Hertz) indicates that the number of repeat waves per second is 100 million.

It is worth mentioning here that some older radio astronomy books refer to the frequency in Cycles per second, or C/S. This is exactly the same as Hertz, so 20 Hz is the same as 20 C/S and 20 MHz is the same as 20 MC/S.

All electromagnetic waves travel at the speed of light, which is 186,200 miles (300,000 km) per second in a vacuum. The keyword here is "vacuum." Electromagnetic waves can slow down, depending on the medium through which they are traveling, for example glass or water.

Electromagnetic waves never travel faster than the speed of light, as this would break the fundamental laws of physics that nothing can travel faster that the speed light, which never happens except in the movies. For the purpose of this book and calculations made within it, we will assume that electromagnetic waves always travel at the speed of light.

The Relationship Between Wavelength and Frequency

If an antenna for a particular frequency is to be made, for example the antennas for picking put Jupiter and the Sun, then the wavelength must be known so the antenna can be cut to size (this will be covered in Chap. 7).

There is a simple equation that can be used to convert frequency to wavelength and vice versa:

$$\text{Wavelength (in metres)} = \frac{300,000}{\text{Frequency (in KHz)}} \text{ (the speed of light in kilometres per second)}$$

$$\text{Frequency (KHz)} = \frac{300,000}{\text{Wavelength (in metres)}} \text{ (the speed of light in kilometres per second)}$$

Polarization of Electromagnetic Waves

An electromagnetic wave is made up of two elements, "electro" and "magnetic," hence the name. The "electro" part of the wave is not the type of electricity we are familiar with, i.e. the flow of electrons through a conductor such as a wire.

Electromagnetic waves do carry the energy of electricity and magnetism, and if they come into contact with a conductor, for example an antenna, they will induce an electrical current through the antenna. This current flow is extremely small, although it would be enough for a suitable receiver to detect and amplify it into a useable signal.

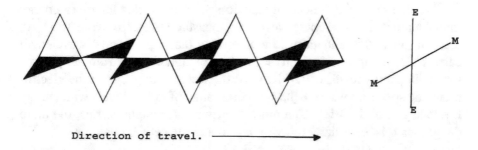

Fig. 2.2 Polarization of electromagnetic waves. An example of vertically polarized electromagnetic waves. Horizontal polarized waves will have the force lines the opposite way around

The electrical and magnetic parts of an electromagnetic wave travel in the same direction but are offset from each other by 90° (see Figure 2.2).

Polarization deals only with the electrical energy within the wave. It is important to know the polarization of the incoming electromagnetic waves to maximize the antenna's ability to receive them. If it were possible to see these waves approaching, and the movement of the electrical energy of the wave was moving up and down, then the magnetic energy would be moving left to right, meaning this wave would be classed as being *vertically polarized*. On the other hand, if the electrical energy of the wave was moving left to right and the magnetic energy was moving up and down, then this wave will be *horizontally polarized*. Manmade signals are usually vertically polarized, while natural radio signals are usually horizontally polarized, although there can be exceptions.

The polarization of an electromagnetic wave may not be fixed, and can rotate continually, sometimes randomly, either clockwise or counter clockwise. This is known as circular polarization. This is especially so if the wave has encountered anything that has caused it to change direction, for example reflection from an ionized meteor trail.

The Electromagnetic Spectrum

We are taught at an early age the color red is associated with hot things and danger, such as stop lights on traffic signals, warning signs etc. The color blue, on the other hand, is associated with cold things. A good example is a sink or bathtub: the red tap is hot water and the blue tap is cold water.

In nature, however, the opposite applies: red stars are cool, with surface temperatures around 3000 K, while blue stars are hot, with surface temperatures around 20,000 K.

The blue end of the electromagnetic spectrum carries far more energy than the red end. These energy levels are measured in electronvolts (eV). An electronvolt is the amount of kinetic energy that a particle or wavelength carries. At the high energy end of the electromagnetic spectrum are gamma rays with energy levels of 124 KeV. At the lower energy end of the electromagnetic spectrum we have the sub-hertz range of radio signals with energy levels as low as 12.4 feV. The more energy a wavelength carries, the more dangerous it is, especially to living tissue and the cells it contains.

Gamma Rays

Gamma rays are the most energetic wavelengths at 10 pm and shorter. These are blocked by the Earth's atmosphere, which is beneficial as they carry so much energy they can smash into living cells and damage the nucleus within the cell, changing its DNA. This can cause it to mutate and possibly become cancerous.

The Apollo astronauts recall seeing bright white flashes in their eyes even with their eyes shut. Gamma rays and other high energy wavelengths can be a real problem to any future astronauts flying to the Moon.

Gamma rays are thought to be emitted by high energy pulsars and super-novae explosions. A strange phenomenon known as a gamma ray burst has been recorded. These bursts are like Lorimer bursts—very short-duration bursts of very high energy levels that seem to come from a point source in space. These are still being researched to find out where and what they are. Several theories have been put forward, such as hypernovae (a super-supernova), but no one knows for sure.

X-Rays

X-rays are divided into two types: hard X-rays and soft X-rays. Hard X-rays have a wavelength of 10 pm–100 pm, and soft X-rays have wavelengths of 100 pm–10 nm.

The longer wavelengths of soft X-rays are the type used by hospitals to find broken bones. They can also be used to cure or prolong the life of some cancer patients by irradiating their tumorous cancers.

The Earth's atmosphere is opaque to this wavelength.

X-ray astronomy began in the late 1940s with the discovery that the Sun emits X-rays. This particular wavelength has been proven to be a great aid in viewing the Sun, as it shows details that are masked in other wavelengths like visible light and hydrogen alpha. X-rays are also a useful wavelength for looking at the Moon. The Moon itself doesn't emit X-rays, but when an X-ray flare is emitted from the Sun it travels through the Solar System, and when it hits the Moon it can cause certain elements within the Moon's surface to glow at X-ray wavelengths. This occurs in a similar way to how white items of clothing seem to glow under an ultraviolet light source. This allows astronomers to study the properties of these elements on the lunar surface and see detail that is not possible at visible wavelengths.

X-rays are also thought to be emitted by black holes. As gases are drawn into orbit around a black hole, the speed at which the gases orbit gets faster and faster. This causes the gases to become superheated to the point where there is so much energy within the gas molecules that they emit X-rays.

Ultraviolet

The ultraviolet part of the spectrum can be subdivided into two categories: extreme ultraviolet, with wavelengths of 10 nm–100 nm, and near-ultraviolet, with wavelengths of 100 nm–350 nm.

The near-ultraviolet wavelength is responsible for suntans. Ultraviolet wavelengths have proven very useful in the study of white dwarfs (a corpse of very dense material remaining after a star with a similar mass to the Sun has ended its life). Ultraviolet has also been used to study the atmospheres of planets, such as Venus and Jupiter. It was one of many wavelengths used to study Jupiter's atmosphere after the comet Shoemaker-Levy 9 slammed into Jupiter's atmosphere in 1994. It has been used to study details in the upper cloud layers of the planet Venus and Saturn's planetary ring system. Hot blue stars also radiate strongly in ultraviolet light. An example of this is M45 the Pleiades, a beautiful open star cluster. Ultraviolet radiation also has an important effect on the Earth's ionosphere, and in turn on the radio astronomy projects discussed in later chapters.

Visible Light

We are all familiar with visible light. Its wavelength is approximately 400 nm–700 nm. As a percentage, the visible light portion of the electro-magnetic spectrum is very small.

At 400 nm, light is violet. The amount that can be seen at this wavelength depends on the individual. It has been found that people are more sensitive to the violet end of the visible light spectrum after a recent cataract opera-tion. The color blue comes in at around 450 nm, followed by green at 500 nm, yellow at 550 nm, orange at 600 nm and finally red at 700 nm.

Electromagnetic radiation is particularly unusual in that it has some of the properties of both a wave and a particle, but not all the qualities of either one. Consider visible light: experiments have been carried out using thin slits cut into a screen to produce a diffraction pattern on a nearby wall. These diffraction patterns never seem to completely fit with what scientists are expecting to see, thereby demonstrating this unique property.

Infrared

The infrared part of the electromagnet spectrum covers quite a large swath, including all wavelengths from 750 nm to 1 mm.

Infrared is subdivided into three groups: near-infrared at a wavelength of 750 nm–2500 nm, mid-infrared at 2.5 μm–10 μm, and far-infrared at 10 μm–1 mm.

The Earth's atmosphere is opaque to much of this wavelength. A limited amount of infrared astronomy can be done from the Earth's surface, but this has to be conducted at high mountains and the environment must be dry, as water vapor in the Earth's atmosphere will absorb some infrared wavelengths.

There are airborne infrared observatories, such as the Stratospheric Observatory for Infrared Astronomy (SOFIA). This is a Boeing 737 that has been modified to carry an infrared telescope. When in flight, a panel at the rear of the plane opens and the telescope can then make observations. It is capable of flying above 99% of the water vapor within the Earth's atmosphere. Infrared astronomy is extremely useful for looking through dust clouds that are opaque to visible light, for example through the plane of the galaxy towards the galactic centre, and for observing star birth in places such as the Orion Nebula.

Submillimetre

The submillimetre wavelength is 0.3 mm–1 mm. This part of the electromagnetic spectrum is where optical detection methods and radio detection methods meet.

The antennas must be made to a higher degree of accuracy than other radio antennas operating at longer wavelengths. The telescopes must be placed at high altitude on top of mountains to get through the thickest part of the atmosphere. The environment must be very dry, as water vapor in the atmosphere can block this wavelength.

Areas such as the Atacama Desert in Chile are suitable. Here, a large submillimetre array is being built on the top of a high-altitude plateau. The radio telescopes are built at lower altitudes and then are moved up the mountain by a specially engineered transport vehicle. Once the radio telescopes reach the plateau, they are then fitted into position. At this altitude, construction workers have to work using breathing apparatus, as the air is noticeably thinner.

This is a new area of research in astronomy. Astronomers are using this wavelength to better understand molecular clouds, star formation and the evolution and formation of galaxies.

Here is a link to a short video of the submillimetre array: https://www.youtube.com/watch?v=yjVVCfDrx8.

Radio

The radio part of the electromagnetic spectrum has very long wavelengths, ranging from 1 mm to 100 km or more. Unlike the other sections of the electromagnetic spectrum, where the frequency at which one section changes to another can sometimes be a little vague (for example, near-infrared and mid-infrared, and also where hard X-rays meet soft X-rays), the radio spectrum is divided into the designated frequencies described below. However, even these can get grouped together, for example very low frequency (VLF) and the four groups below it are sometimes treated as a large subgroup and collectively referred to as very low frequency (VLF).

Name: Terahertz
Abbreviation: THz
Frequency: 300–3000 GHz
Wavelength: 1–0.1 mm
Examples: This is still experimental, but could prove a replacement for X-ray imaging where normal X-ray equipment could not be used because of access or because the area of the body on which it is going to be used could be damaged by X-rays. It also has scientific uses, such as higher frequency processors to enable faster computing and the relatively new field of submillimetre astronomy.

Name: Extremely high frequency
Abbreviation: EHF
Frequency: 30–300 GHz
Wavelength: 10–1 mm
Examples: Microwave remote sensing and microwave radio relays.

Name: Super high frequency
Abbreviation: SHF
Frequency: 3–30 GHz
Wavelength: 100–10 mm
Examples: Modern radar, microwave devices.

Name: Ultra high frequency
Abbreviation: UHF
Frequency: 300–3000 MHz
Wavelength: 1 m–100 mm
Examples: Television broadcasts, cellphones, GPS, Bluetooth devices.

Name: Very high frequency
Abbreviation: VHF
Frequency: 30–300 MHz
Wavelength: 10–1 m
Examples: FM radio, ground to aircraft and aircraft to aircraft communication, maritime communication.

Name: High frequency
Abbreviation: HF
Frequency: 3–30 MHz
Wavelength: 100–10 m
Examples: Shortwave broadcasts, amateur radio and other commercial radio broadcasting and communication. The 20 MHz range is used to receive radio emissions from the Sun and Jupiter.

Name: Medium frequency
Abbreviation: MF
Frequency: 300–3000 KHz
Wavelength: 1 km–100 m
Examples: Medium-wave radio broadcasting.

Name: Low frequency
Abbreviation: LF
Frequency: 30–300 KHz
Wavelength: 10–1 km
Examples: Long-wave radio broadcasting, navigation and timing signals.

Name: Very low frequency
Abbreviation: VLF
Frequency: 3–30 KHz
Wavelength: 100–10 km
Examples: Natural radio signals, communications with submarines and other military communication, emergency beacons, e.g. avalanche beacons.

Name: Ultra low frequency
Abbreviation: ULF
Frequency: 300–3000 Hz
Wavelength: 1000–100 km

Examples: Natural radio signals, communication within mines.

Name: Super low frequency
Abbreviation: SLF
Frequency: 30–300 Hz
Wavelength: 10,000–1000 km
Examples: Communications with submarines, mains/grid AC electrical supply (50–60 Hz) and natural radio signals.

Name: Extremely low frequency
Abbreviation: ELF
Frequency: 3–30 Hz
Wavelength: 100,000–10,000 km
Examples: Brain electrical activity, communication with submarines, electromagnetic waves from the Earth's ionosphere.

Name: Sub-hertz
Abbreviation: sub-Hz
Frequency: <3 Hz
Wavelength: >100,000 km
Examples: Geomagnetic pulsations, electromagnetic waves from the Earth's ionosphere, and possibly earthquakes. Researchers studying earthquakes are looking into the possibility that the detection of sub-hertz and other very low frequency waves could be used as a precursor to an earthquake, as very low frequency waves are given off by the underground strata as the stresses build up within the strata before the ground moves to produce an earthquake.

Properties of Different Wavelengths and Frequencies

No matter what the wavelength, radio waves have the same characteristics, albeit with different laws of propagation.

Higher frequencies are more easily reflected from large objects such as hills or large buildings than lower frequencies. If they come into contact with an obstruction, for example a large building that is greater in size than the wavelength of the wave, they will be reflected like a light beam hitting a mirror. Higher frequency radio waves have a tendency to behave like light waves due to their shorter wavelength.

Lower frequencies with their longer wavelengths have the ability to travel around or over such objects. For example, a frequency of 3 Hz that has a wavelength of 100,000 km could easily pass an object the size of the planet Earth without any trouble.

Each frequency range has its own unique qualities. The higher frequencies are useful for sending large amounts of information and as a rule are received by relatively small antennas. The lower frequencies have good penetrating properties and can travel through the deepest oceans for communication with submarines, and have antennas measured in kilometers.

Radio astronomers have a number of protected frequencies (see Chap. 19 that are left clear from traffic so that they can receive the weak signals from space without too much interference. One of these protected frequencies is the 21-cm hydrogen line, mentioned earlier.

Thermal Radiation

When a radio telescope is used to receive radio energy from an extraterrestrial source, radio astronomers are interested in two types of radio energy or radiation: thermal radiation and synchrotron radiation.

Thermal radiation is electromagnetic energy that is given off by a hot object. If we use the spectral classification of stars to illustrate this, blue stars are far hotter than red stars. B class stars can range in surface temperature from 10,000 K to 30,000 K, while on the other hand a red M class star has a surface temperature range of 2500–3900 K. If we set aside the fact that some red stars are huge, as in the case of Betelgeuse, and just stick to the surface temperature, blue stars will emit more thermal energy than their red counter parts.

This means that there is a direct correlation between color and temperature. An atom has a certain energy level to start with—this is known as its *ground state*. If energy—for example an electrical charge—is applied to that atom or to a molecule (a molecule being a combination of two atoms or more), the electron(s) that are orbiting the nucleus of the atom(s) will become excited, causing them to jump to a higher energy state or orbit.

Likening the Solar System to an atom, with the Sun as the nucleus and the planets as orbiting electrons, the effect of the electron jumping to a higher energy state or orbit would be like the Earth jumping out to the orbit of Mars. This jump to a higher energy state or orbit is just that—a jump, and not a slow migration. The energy that is applied to the atom(s) can be steadily increased, but no change will take place until the exact amount of energy needed to make the electron(s) jump to a higher energy state is applied.

If an atom loses energy, then the electron(s) jump back to a lower energy state or an orbit, closer to the nucleus of the atom. When this happens a photon will be released, and this will have its own unique wavelength.

A good example of this phenomenon, and one that is quite easy to do, is to place a piece of copper, such as a soldering iron used to solder sheet metal (not the type used to solder electrical components), into a gas flame to heat it. When the soldering iron reaches the right temperature for soldering, the gas flame turns a beautiful emerald green color. This is due to the atoms of copper releasing photons at the wavelength that the human eye sees as the color green.

This is the same reason why images of the Sun taken in hydrogen alpha light are always a red-orange color, because this is the wavelength of hydrogen alpha.

This is fine for something that can be seen. Take a piece of metal and with a oxy-acetylene torch, heat up the metal until it starts to glow a dull cherry red. Here, photons of the color red are being released. If the heating is stopped, the metal will start to cool after a few minutes, and the dull cherry red color will disappear. The metal will look visually the same as it did before it was heated, but it will still be hot enough to burn the hand if it is touched. Heat radiating from the metal can still be felt. This is infrared radiation, but because our eyes are not sensitive to infrared radiation, it cannot be seen.

In space, the average temperature is 2 or 3 degrees above absolute zero. If this experiment was conducted again in space, even if the metal has cooled below the temperature that can be detected within the infrared part of the spectrum, the metal will still be radiating energy as long has its temperature is above the average temperature of space. This energy will be radiating beyond the longer wavelengths of the radio part of the electromagnetic spectrum.

While studying radio astronomy, the term "black body" or "black body radiation" may crop up, especially in books dealing with radio astronomy and spectroscopy. This may seem a little complicated, but simply put: a *black body* is a theoretical object that absorbs thermal energy (heat) and is a perfect emitter of thermal radiation. The characteristic of thermal radiation is that it has a continuous spectrum. The best black body in the Solar System is the Sun. If we could stand at the side of the Sun with a blowtorch, the Sun would easily absorb the heat from the blow torch and it is an almost perfect emitter of thermal radiation.

Thermal radiation was causing the hissing noise that Robert Wilson and Arno Penzias received with their horn antenna. This was the thermal radiation left over from the Big Bang at the start of the universe, which has now cooled from the unimaginable temperature at the moment of the big bang down to 2 or 3 degrees above absolute zero over the last 13.8 billion years.

Synchrotron Radiation

The next type of electromagnet radiation that radio astronomers are interested in receiving is synchrotron radiation, so named because it was first observed in particle accelerators called synchrotrons.

The characteristics of synchrotron radiation compared to that of thermal radiation are quite different and make synchrotron radiation easier to detect.

Synchrotron radiation is produced by particles, usually electrons, accelerated to high speeds within a magnetic field. This may sound familiar, as it was used when searching for the Higgs boson at CERN using the Large Hadron Collider (LHC). The LHC used high-powered magnetic fields to accelerate particles up to 99% of the speed of light and smashing them into each other, then imaging the particles created by these collisions.

The planet Jupiter with its vast and powerful magnetic field would make a very good particle accelerator, but it may prove a little bit difficult to control.

Imagine an electron spiraling along a magnetic field line within a magnetic field. As the electron spirals along the field line, it produces electromagnetic radiation. The frequency of the emission is directly related to how fast the electron spirals: the faster the spiral, the higher the frequency. If a stronger magnetic field is used, this will tighten the spiral for the electron and therefore increase the frequency of the electromagnetic radiation produced. Due to this spiraling motion, synchrotron radiation emissions are polarized.

In radio astronomy, synchrotron radiation comes from some of the most powerful sources in the universe. M87 in the constellation of Virgo is a good example of a radio galaxy. This radiation has also been detected from quasars and supernova remnants, such as the Crab Nebula, and in particular the high energy particles coming from the central pulsar and Jupiter's powerful magnetic field.

There are a number of podcast about the electromagnetic spectrum available from www.astronomycast.org:

- Episode 130: Radio Astronomy
- Episode 131: Submillimetre Astronomy
- Episode 132: Infrared Astronomy
- Episode 133: Optical Astronomy
- Episode 134: Ultraviolet Astronomy
- Episode 135: X-ray Astronomy

Chapter 3

The Audio Spectrum

Sound

The age-old question is: If a tree falls in the woods and there is no one around to hear it, does it make a sound?

The answer to this question is a resounding yes!

Science has defined sound as compression waves traveling through a medium. The medium can be a gas, liquid or a solid. So even if there is no one there to hear the tree fall, the compression waves will still be generated as the trunk of the tree begins to break and then as the tree hits the ground.

Sound waves traveling through a gas or a liquid have roughly the same qualities, although sound travels further and faster in water than in air. Sound traveling through a solid such as rock (in the case of an earthquake) has slightly different properties of propagation, but for the following examples this difference can be ignored.

Sound is made up of compression waves and cannot travel through the vacuum of space, as there are no molecules to carry the sound. It has been somewhat overlooked when sending out space probes to visit planets within the Solar System.

The only probe that has had a microphone fitted to it was the Huygens probe. The Cassini spacecraft dropped the Huygens probe on its way to Saturn. Huygens landed on Titan, a moon of Saturn, on January 14, 2005.

As the probe made its descent through the atmosphere of Titan, the microphone was switched on, and for the first time we heard sound on another world other than the Earth and Moon.

Here is a link to the real audio as the Huygens made its descent: https://www.youtube.com/watch?v=1Dyw0dVc.

Meanwhile, sound has now become quite an important area of study within astronomical research, especially the realm of SETI.

Take for example SETI searches looking for radio signals from intelligent life. There are two very good reasons that intelligent life should be able to hear sound waves in one form or another. Almost all multi-celled life on Earth can hear or sense sound waves. This ability is needed to find food and to avoid being food for other animals. Bats for instance make incredible use of high frequency sound for echo location. Life here evolved to suit the environment of the Earth, including the temperature and pressure both of the atmosphere and the oceans. Humans can hear frequencies between 20 Hz and 20 KHz. Our ability to hear the higher frequencies is lost with age or from working in a noisy environment.

On other planets, sound waves would travel differently and therefore any inhabitants of that planet would hear sound waves differently. A quick look at our own Solar System and its planets will show how things may be radically different to how sound works on the Earth.

In the following scenarios, we shall ignore any other environment conditions such radiation, heat, cold and pressure, and assume that our two astronauts remain unharmed.

The Sun

The Sun's atmosphere, the corona, is over a million degrees in temperature. The atoms within the corona are widely spaced apart and traveling at very high velocities. In this first scenario, the Sun's corona will be our medium for the sound waves to travel. If our two astronauts were trying to talk to each other, the molecules within the corona are so reified that even if they shouted to each other, it wouldn't be more than a whisper. If we now take into account the Sun's corona, which is almost all hydrogen, the pitch of their voices would change similar to someone who has breathed gas from a helium balloon (this is *not* recommended to try as it can be deadly), but even more so since hydrogen is less dense than helium.

The astronauts would be able to hear solar activity from solar flares and other phenomena, but this would still be affected by the same distortion as their voices.

Mercury

Mercury has been found to have a very tenuous atmosphere. If Mercury's atmosphere was replicated on Earth, it would be something very close to a vacuum. The first thing the two astronauts standing on the surface of Mercury would notice is how quiet it is. The tenuous atmosphere doesn't have enough molecules to carry any sound waves. The only medium on Mercury capable of carrying sound waves is the planet itself. The two astronauts would only hear anything if they placed their ears in contact with the surface rock, and the only things they are likely to hear are Mercury quakes within the planet.

Venus

Venus's atmosphere is almost all carbon dioxide, and at a pressure of around 90 times that of the Earth at sea level, its atmosphere is hot at around 842 °F (450 °C). So, the atmosphere is our medium in which sound waves can travel. If we consider the atmospheric pressure and temperature, Venus's atmosphere behaves more like a liquid than a gas.

Our two astronauts standing on the surface of Venus would experience sound in a way similar to how sound is heard underwater on the Earth. The astronauts would hear sounds from Venus, like lightning discharges and volcanic activity, but this too would sound distorted.

Mars

Mars's atmosphere is almost all carbon dioxide, with a few trace elements throw in. The atmospheric pressure on Mars is low and only enough to blow fine dust around. This fine dust can stay suspended in atmosphere for weeks and engulf the entire planet in a global dust storm, blocking all views of the surface. Global dust storms have a nasty habit of occurring when the planet is at its closest approach to Earth, frustrating optical astronomers. This is thought to happen because as the planet gets closer to the Sun, the heating of the atmosphere changes its density, which increases the likelihood of such dust storms.

There are sufficient molecules within the atmosphere to carry sound, although this can be likened to the upper limits of the atmosphere here on Earth. Our two astronauts on the surface of Mars would probably hear the sound of their feet as they walked about on the surface. If they tried to speak they would hear each other, but only as if they were whispering.

Jupiter and Saturn

Jupiter, the largest of all the planets, is made up of gas. Traveling down through the atmosphere, the density of the gases increases and so too does the pressure, so much so that the gases eventually turn it into a liquid.

If we were to stay within the first few hundred miles of the clouds, our two astronauts may find it not dissimilar to the Earth. However, Jupiter has far greater lightning discharges, many hundreds of times more powerful than the Earth. The sound of the lightning discharges would be deafening and travel through the atmosphere with great force, and their echoes would ricochet like a starting pistol being fired inside a railway tunnel.

Saturn, being of a similar structure to Jupiter, would behave and sound in a similar sort of way as Jupiter. The lightning discharges would be slightly less powerful.

Uranus and Neptune

Uranus and Neptune's atmospheres are mostly hydrogen and helium. So, a human voice would sound very high pitched, but Uranus and Neptune are very cold worlds. The super cold temperature of the atmosphere would bring down the speed of sound and drop the pitch slightly, but not enough to stop our two astronauts sounding like cartoon characters.

Why the Audio Spectrum Is Important to the Radio Astronomer

A radio telescope collects radio waves and produces a signal. Quite often, this signal is turned into audible sounds we can hear and try to understand. A radio astronomer needs to have a good understanding of the audio spectrum (20 Hz–20 KHz) as well as the radio spectrum. This allows the radio astronomer to spot signals sometimes hidden within the audio spectrum, and to use different filters to help isolate the audio we want. This is done by using software to repress some frequencies and boost others. This is not unlike optical astronomy: use the right filter on the right object and the view is enhanced, but use the wrong filter and nothing will be seen.

Here is a guide to audio frequencies and how changing each frequency changes the audio:

- **20–60 Hz**: This frequency range is sometimes called the "muddy" range. Interference from mains power supplies falls within the 50–60 Hz range, so reducing the 50–60 Hz range can help remove "mains hum." Software such as Spectran and Spectrum Lab come with built-in hum filters. Its main use is for low base frequencies. Don't bother too much with this frequency unless interested in low frequency radio astronomy.
- **60–250 Hz**: These frequencies are said to add warmth to audio and voice recordings. Too little and the voice will sound dull and flat; too much and the voice will sound loud and booming. This range can be useful in trying to isolate an astronaut's voice from background noise.
- **250 Hz to 2 KHz**: These frequencies make things sound "tinny," like cheap headphones. Dropping these frequencies can help remove some of these tinny sounds.
- **2–7 KHz**: These frequencies carry a lot of detail. Boosting these frequencies can help the listener hear subtle detail that may otherwise be overlooked.
- **7–15 KHz**: Like the previous range but to a lesser extent, these frequencies can be boosted to help the listener hear subtle detail that may otherwise be missed.
- **15–20 KHz**: Don't bother too much with these frequencies, and if over 30 years of age, chances are you won't be able to hear them anyway.

Chapter 4

The Earth

The Earth's Atmosphere

A planet's atmosphere is a very important part of its makeup. Sometimes it can be overlooked or taken for granted, especially on the Earth.

To understand the scale and depth of the atmosphere on Earth: if we were to reduce the Earth's size to that of a soccer ball, dip the soccer ball in a bowl of water, then lift it out, the layer of water left on the ball would be the same as the Earth's atmosphere.

Looking at the four terrestrial planets, Mercury, Venus, Earth and Mars, we see a vast difference in the atmosphere of each.

It was first thought that Mercury didn't have an atmosphere, but thanks to the MESSENGER (MErcury Surface, Space ENvironment, GEochemistry, and Ranging) space probe, Mercury has been found to have a very thin atmosphere. The planet gets baked on the sunlit side and frozen on the dark side. The Sun has irradiated Mercury's surface with lethal doses of radiation and charged particles.

The planet Venus has a very dense atmosphere of almost pure carbon dioxide. The atmospheric pressure is 90 times that of the Earth. The heat at the surface is hot enough to melt lead, not to mention the sulfuric acid rain!

Mars's atmosphere is mostly carbon dioxide, but so thin it barely exists. Mars is so cold that carbon dioxide freezes out of the atmosphere in its winter and collects at the poles.

© The Editor(s) (if applicable) and The Author(s), under exclusive
license to Springer Nature Switzerland AG 2021
S. Arnold, *Radio and Radar Astronomy Projects for Beginners*, The Patrick
Moore Practical Astronomy Series, https://doi.org/10.1007/978-3-030-54906-0_4

The Earth is in the *Goldilocks Zone*—not too hot and not too cold, but just right for habitability. The Earth's atmosphere is a mixture of approximately 78% nitrogen, 21% oxygen and the last 1% a mixture of carbon dioxide and other gases. Without the atmosphere there would be no life on Earth. It has the ability to protect us from the vacuum and extreme cold of space, as well as the unimaginable heat from the Sun.

Earth's atmosphere is made up of a number of layers, with transition layers between each. The major layers are as follows. (All altitudes are approximate as they can vary from season to season and with the latitude of the observer on the Earth.)

1. The first layer is the *troposphere*. It starts at sea level and rises to 7.5 miles (12 km) in altitude. This is the thickest part of the atmosphere and where the air pressure is at its greatest. Most of the weather occurs here. This is what we call home.
2. The *stratosphere* is the second next major layer and covers the range from 7.5 miles (12 km) to 30 miles (50 km). This layer contains the ozone layer at 12.5 miles (20 km) to 24.8 miles (40 km).
3. The *mesosphere* is the third major layer. This starts at 31 miles (50 km) and ends at 52.8 miles (85 km). This is the altitude at which the rare noctilucent clouds are seen and were meteors burn up. It is also the start of the ionosphere, which will affect our radio astronomy projects.
4. The *thermosphere* is the fourth major layer, starting at 52.8 miles (85 km). Due to solar activity, its ending fluctuates in altitude between approximately 217 miles (350 km) and 466 miles (750 km).
5. The *heterosphere* is the fifth major layer. This is where the gases in the atmosphere start to separate into component parts and the mixture that we know as air is lost. The lighter trace gases such as helium and hydrogen simply float away.
6. The last major layer is the *exosphere*. This is the outer most layer of the atmosphere and the realm of the two gases hydrogen and helium. The gases at this altitude are so rarefied that they barely interact with each other.

As a percentage, the atmosphere doesn't add up to much, but life on Earth would be in trouble without it. The atmosphere plays a large part in blocking certain wavelengths from getting through to the Earth's surface. Figure 4.1 shows the effectiveness of the atmosphere in stopping these harmful wavelengths.

Earth's atmosphere blocks everything from gamma rays right up to the longer wavelengths of ultraviolet, just before the visible wavelengths. It is transparent to visible light but starts to become opaque in the infrared, with only a few wavelengths getting through. The longer wavelengths of the far infrared are blocked, and then it becomes transparent for parts of the radio wavelengths.

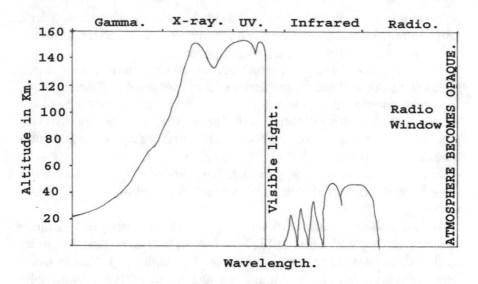

Fig. 4.1 This image shows the effectiveness of the atmosphere at blocking short wavelengths of electromagnetic radiation

Looking at the altitude at which each wavelength is blocked it, we can see why satellites and space telescopes are needed to effectively observe wavelengths such as ultraviolet and X-ray. As we have discussed previously, some useful astronomy can be done with high altitude balloons for the X-ray wavelengths, and some useful infrared astronomy can be done from the ground, but water vapor in the atmosphere blocks a lot of infrared wavelengths. This is why airborne infrared telescopes are used, as they can fly at an altitude where they are above 99% of the water vapor in the atmosphere.

The Ionosphere

As mentioned earlier, radio waves were once thought to travel in straight lines. This is known as a line-of-sight phenomenon.

When Marconi made his trans-Atlantic transmission in 1900, it was realized that some sort of reflection must be taking place to get the transmission around the curvature of the Earth. In the mid-1920s, experiments were carried out that proved the existence of the ionosphere, and a mathematical model was put forward. The big leap came with the first artificial satellite Sputnik 1, launched in 1957. This satellite, broadcasting its "beep, beep" signal at 20 MHz and 40 MHz, could be heard when the satellite orbited overhead. This proved without question that the ionosphere could be

penetrated if the right frequency was used. The launch of Sputnik 2 with
Laika on board further narrowed down the frequencies that scientists identified were able to travel through the ionosphere.

The ionosphere is a very important factor in some of the projects covered
in this book. We will use the properties of the ionosphere in different ways,
so a good knowledge of it would be useful. Radio enthusiasts around the
world use the reflective qualities of the ionosphere to bounce their signals
great distances to other radio enthusiasts, in a practice known as DX communication or *DXing*.

A vast amount of work as gone into the study of the ionosphere and a
large amount has been written on the subject. What follows is a very basic
summary.

The *ionosphere* is a shell that surrounds the Earth. It starts at an altitude
of approximately 43.5 miles (70 km) and extends to more than 373 miles
(600 km). It owes its existence and variability to the high energy wavelengths from the Sun, such as ultraviolet and X-ray. It changes with solar
activity as well as the Sun's 11-year solar cycle from solar maximum to solar
minimum. In particular, activity such as solar flares can produce a marked
effect on the ionosphere. There are also seasonal variations to the ionosphere
for an observer on the Earth, who will notice a change from midsummer to
midwinter as the height of the Sun changes in altitude from month to month.

The ionosphere can change over a few hours from day to night. A good
way to demonstrate just how quickly it can change is to look at Fig. 4.2. The

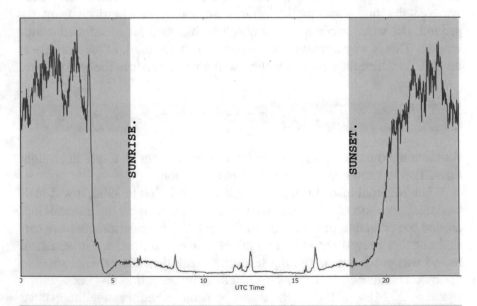

Fig. 4.2 A day of SuperSID data used to show changes to the ionosphere

image is of a day's worth of SuperSID data. The SuperSID monitors the ionosphere and uses changes within it to pick up solar activity such as X-ray flares. There is no need at this stage to understand the graph, as we are only interested in the change to the ionosphere shown by the graph.

At night, there is very little ionization. This can be seen by the height of the graph in the nighttime hours. Looking at the sunrise and sunset sides, which have been marked on the graph, with time across the bottom axis, it can be seen that the ionization of the ionosphere at sunrise happens almost instantly, as demonstrated by the near-vertical line on the sunrise side. On the sunset side, the line is at an angle, indicating that it takes a little longer (a couple of hours or so) to change and lose the ionization.

The ionosphere is made up of a number of layers. The only layers that are of interest to us are the D layer, E layer and F layer. Each layer is ionized by different types of radiation. The inner D layer is affected by hard X-rays, the E layer with soft X-rays and extreme ultraviolet, while the outer F layer is ionized with extreme ultraviolet and near ultraviolet. During the day, the side of the Earth facing the Sun is fully exposed to the radiation emitted by it. This causes a large increase in the ionization of the atoms in the D, E and F layers. The outer most F layer separates into two layers, F1 and F2. See Fig. 4.3.

This separation of the F1 and F2 layers during the daytime allows for long-distance communications at high frequencies, as radio waves can be reflected greater distances by the higher F2 layer. The E layer is accessible to wavelengths of 80 m to 2 m, and the D layer in the daytime absorbs wavelengths of between 160 m and 40 m.

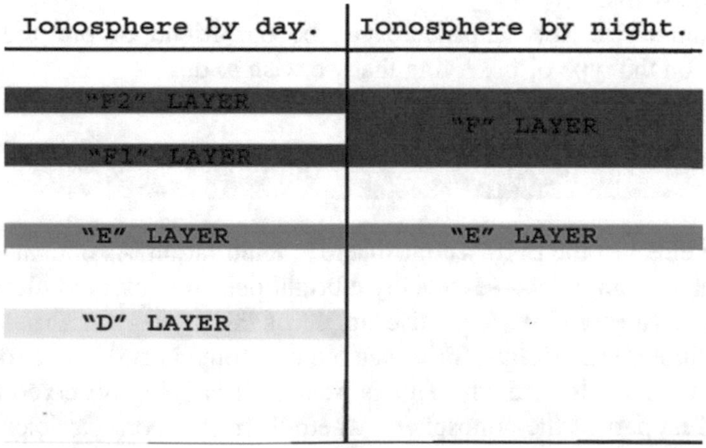

Ionosphere by day.	Ionosphere by night.
"F2" LAYER	
"F1" LAYER	"F" LAYER
"E" LAYER	"E" LAYER
"D" LAYER	

Fig. 4.3 The change in the layers of the ionosphere by day and night

It can be seen that the effects of this ionization have a lot to do with the wavelengths that are being used. The ionization is excellent news for the SuperSID monitor, as it relies on the reflective qualities of the ionosphere to function. However, this ionization is bad news if trying to pick up Jupiter, as it makes it almost impossible during the day. As a rule, the signals from the planet Jupiter hit the ionization and are reflected back into space and lost. The ionization is less of a problem for the radio emissions from the Sun, as they are strong enough to force their way through, and this allows them to be received.

The ionization also affects the Very Low Frequency (VLF) INSPIRE receiver. It traps the very low frequency signals between the Earth and the ionosphere and bounces them between the two, stopping the signals traveling higher than this ionized layer and limiting the types of sounds that can be heard.

As mentioned, the level of ionization within the ionosphere can change. A couple of hours or so after sunset, the ionosphere can "cool down;" the two F layers recombine as they lose some of their ionization. The E layer also loses most of its ionization, and the D layer can disappear completely. In the case of the SuperSID monitor, when the sun sets, the monitor ceases to function, as it relies on the reflective quality of the ionosphere to work. This is the time to pick up the radio emissions from the planet Jupiter, as these radio emissions can now pass through the ionosphere to the receivers.

In the height of summer, or if solar activity is high (maybe there are a large number of sunspots on the Sun), the ionosphere can remain partly ionized all night. This makes receiving Jupiter's radio emissions very difficult if not impossible.

To summarize, the ionosphere can be our friend or our foe—it all depends on the type of observing that we wish to do.

Scintillation

Another effect of the Earth's atmosphere is scintillation. As optical astronomers know, if an object—especially a bright one—is observed close to the horizon, it can twinkle. A good example of this is the star Sirius. When viewed through binoculars, Sirius can seem to come in and out of focus and can change its color quickly. This is because it is being observed through the thickest part of the atmosphere, where there are more particles of dust and pollution to contend with. All of these factors play a part "seeing" an object. As a general rule, all objects within 15–20° of the local horizon are prone to this effect unless the "seeing" is very good. The higher an object is in relation to the observer's horizon, the better.

This is also true of radio astronomy. The radio waves are traveling parallel to each other as they head towards the Earth's atmosphere. As they travel through the ionosphere, the ionization can be irregular; it may be heavier in some areas and less in others. Thus, the radio waves leaving the ionosphere can be affected by the irregular ionization in the same way as light from a star seems to make it twinkle. Radio emissions from an object, for example the planet Jupiter, are less troubled if the height and angle of the planet in relation to the observer is greater, as the antenna is less likely to pick up interference and other stray signals.

The Earth's Magnetic Field and Magnetosphere

A planet's magnetic field plays an important role in radio observations, especially those of the Earth and Jupiter. Looking at the four terrestrial planets and comparing their magnetic fields, we shall see how the Earth is quite unique in having a strong magnetic field. But even this is dwarfed by the gas giant Jupiter.

Mercury has a small magnetic field despite its size. Its magnetic field is only around 0.01 times that of the Earth's. The magnetic field is thought to come from a large nickel-iron core. This core is now thought to be solid, as Mercury is too small for it to have hung onto its internal heat. This magnetic field, although weak, is able to stop some of the charged partials from the solar wind, but the lack of any substantial atmosphere means that other particles can get through to the planet's surface and bathes Mercury in lethal doses of radiation.

Venus is a real mystery. A planet roughly the same size as Earth, Venus should have a magnetic field similar to that of the Earth, but in reality, it has very little to none. Space probes haven't managed to measure a magnetic field yet, it is that weak! The best estimates are that Venus does have a magnetic field, but it is in the order of 0.00001 times that of the Earth.

Venus is large enough to have kept its internal heat and still has volcanic activity, but because Venus rotates very slowly (243 Earth days) it is thought that there isn't the same dynamo effect as on the Earth. Without a protective magnetic field, Venus's atmosphere is slowly being lost to space due to the action of the solar wind. Space probes that have ventured inward towards Mercury and Venus have observed a stream of gases coming from Venus in a similar way to the stream of ices and gases that produce the tail of comets.

Mars has a magnetic field, but it's only 0.002 times that of the Earth. Mars rotates fast enough (just over 24 h) to have a dynamo effect, but because of its small size, it has lost its internal heat, resulting in the solidification of the core.

The Earth's magnetic field is far more powerful than the other three terrestrial planets. The magnetic field of the Earth can be likened to that of a bar magnet. The magnetic north and south poles are offset to the rotation of the Earth's axis by approximately 12°. This can change very slightly annually and after large earthquakes. Our magnetic field has flipped in the past, just as the Sun flips its magnetic field every 11 years or so. Samples taken from old lava flows have been examined, wherein small pieces of magnetic rock within the lava lined themselves up with the Earth's magnetic field at the time the lava solidified. This alignment has been found to be the opposite polarity to Earth's magnetic field today. No one knows exactly how or why this process happens.

It would be interesting to know how birds would cope with the reversing of the magnetic field. Migratory birds have magnetic receptor cells within their brains that line up with the magnetic field lines of the Earth, allowing them to navigate the thousands of kilometers between their breeding grounds. A common garden bird—the robin—can go one better. Tests have shown that robins can see the Earth's magnetic field, but only in one eye.

The origin of the magnetic field lies within the Earth's core. The inner core, a solid chunk of iron, floats inside the outer core, a liquid mixture of nickel and iron. The liquid outer core is thought to have the consistency of thick treacle. Convection currents within the liquid outer core are generated as the hot material rises, then drops back down in a circular motion as the material cools. A good way to imagine this is to think of the movement of wax is a lava lamp. The heat from the electric lightbulb heats the wax, and the wax then rises up through the oil in the lamp. As the wax cools, it sinks back down again, and the process repeats. The inner core rotating within the liquid outer core contributes to these convection currents, and combined with the speed of the rotation of the Earth acts like a large dynamo that produces the vast magnetic field of the Earth, and in turn, the magnetosphere.

The Earth's magnetosphere extends many thousands of Kilometers into space, and its distance varies greatly due to its interaction with the solar wind. It contains within it two large torus-shaped areas. These are like two huge tire-shaped zones that surround the Earth. There is an inner and outer torus: these are the *Van Allen radiation belts*, named after James A. Van Allen, who first theorized their existence.

The Van Allen belts are made up of highly charged plasma that has been energized by charge particles from cosmic rays, but mainly from the charged particles contained within the solar wind. These particles then become trapped within the Earth's magnetic field. The outer belt contains mainly electrons and the inner belt contains the higher energy particles. They stop the highly charged particles from reaching the Earth's surface.

The magnetosphere acts like a large buffer against the solar wind. Charged particles from the Sun hit the outer belt and are channeled along the field lines away from the Earth. In times of high solar activity, the incoming pressure of the solar wind on the Sun side can increase greatly and cause deformation in the outer belt. The side away from the Sun can develop a long magnetic tail that can stretch many thousands of kilometers into space. As more and more energy from the solar wind hits the leading edge of the magnetosphere, the tension builds up in the long magnetic tail like an elastic band. At some point the elastic band snaps back, bringing charged particles careening along the field lines and funneling them down at the poles, where they interact with the atoms in the atmosphere. This produces the aurora.

If a large enough amount of energy hits the magnetosphere, such as an X-ray flare from the Sun, the magnetosphere would deform and in turn distort the ionosphere. This is known as a *sudden ionospheric disturbance* (SID), and these are the signals picked up by the SuperSID monitor discussed in Chap. 10.

If low frequency radio waves get through the ionosphere, they get trapped within the magnetic field lines and can travel hundreds of kilometers into space before being brought back down the magnetic field line. This allows them to travel great distances from one hemisphere to the next. These are known as whistlers and are the same signals picked up by Marconi and Tesla when attempting to listen for signals from space. They are also the signals that the INSPIRE receiver picks up and will be discussed in Chap. 10.

Chapter 5

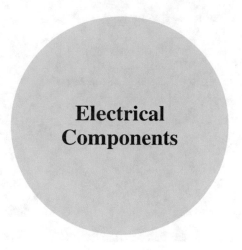

Electrical Components

Electrical Safety

Electricity can kill! If in any doubt about mains/grid power supply, consult a qualified electrician.

If main/grid power is to be used instead of batteries, seek the advice of a qualified electrician if in any doubt.

Please follow any instructions supplied with the kits carefully, especially where polarity is involved. Some electronic components are polarity sensitive. As a rule of thumb: check twice, solder once.

Soldering Irons and Soldering

Solder1ing Irons

There are a bewildering number of soldering irons available in the marketplace today. Here are a few pointers to help make a good purchase.

Try to avoid the cheaper unregulated ones, as these can overheat and burn out. Also avoid the ones that run off of a 12-V car battery. Choose a thermostatically controlled one, as these automatically switch on and off after reaching the correct temperature.

© The Editor(s) (if applicable) and The Author(s), under exclusive license to Springer Nature Switzerland AG 2021
S. Arnold, *Radio and Radar Astronomy Projects for Beginners*, The Patrick Moore Practical Astronomy Series, https://doi.org/10.1007/978-3-030-54906-0_5

Fig. 5.1 A modern, digitally controlled soldering iron

A soldering iron's power is measured in Watts. A good choice would be one around 25–35 W. This would be adequate for all the projects in this book, with the exception of the construction of the Radio Jove antennas, which need an iron of around 75 W to heat the heavier gauge wire that is used.

Most good soldering irons come with the ability to change the tip. The tips come in many different sizes and shapes. A good choose of tip would be one of around 2.5–3 mm (0.125 in.) with an angled end rather than a point. One important accessory is a stand, which holds the hot soldering iron when not in use. These have a little sponge in the base that is kept damp.

Before each new soldering joint, give the iron tip a quick wipe, as this removes any scum off the tip before making a new joint. Don't waste money buying the replacement sponges—just cut a piece off a dishwashing sponge, as it works just as well.

Figure 5.1 shows a modern, digitally controlled soldering iron. The temperature can be adjusted to suit the job at hand.

Tinning a Soldering Iron

It is a good idea at the start of each soldering session to allow the iron to reach the correct temperature. Then, give the tip a quick wipe with the damp sponge and apply a small amount of solder to the tip. The flux in the solder cleans and freshens up the tip before use. This needs to be done only once per use.

Types of Solder

There are a number of different types of solder. Always choose one with a non-corrosive flux, as the acid flux will corrode the joints over time. Solders used to contain large amounts of lead, but modern-day solders are lead-free and come in a number of different diameters: 0.8 mm and 1 mm (0.040 in.) are useful sizes.

How to Solder

Good soldering skills are essential for electrical work. A poorly soldered joint can make the difference between a project that works and one that does not. The first thing to remember is to keep everything clean and free from any grease, oil or melted insulation, which will stop the solder from doing it's job. Any dirt and grease can be removed with a solvent cleaner or a light sanding from a piece of very fine sandpaper. In some books, steel wool is suggested, but if this is used and any fibers of the steel wool go adrift, there is the potential for a short circuit.

The next important thing to remember is: never melt the solder with the soldering iron. Place the soldering iron on the joint for a few seconds and let the heat from the joint melt the solder. Keep the soldering iron there for a few seconds and allow the solder to fill the joint, then remove the soldering iron and allow the joint to cool.

A lot can be learned from the appearance of a soldered joint. It should look smooth and shiny, almost wet. If it looks dull or full of holes, it's probably a bad joint. This can be caused by being too eager and not letting the soldering iron reach the correct temperature before using it, by some sort of contamination within the joint or on the soldering iron, or even by a bad batch of solder.

The best way to learn how to solder is to have a go. A good place to start would be to get some "stripboard" from a local electronics supplier, as this is ideal to practice the art of soldering. Stripboard has pre-drilled holes in it ready for components to be soldered in place. It is insulated on one side and has lines of copper tracks running its full length on the other. It is used as a base for building amateur electronic projects. Also needed is solid core wire of a diameter of about 0.6 mm (0.024 in.).

Strip the insulation off the wire to about 300 mm (12 in.) and cut the bare wire into pieces of about 30 mm (1.25 in.). Bend the last 6 mm (0.25 in.) over so that it looks like a big staple. Place these pieces through the holes on the stripboard so that the ends of the wire appear on the side where the copper tracks are. Now try and solder this into place without getting solder on the other tracks from where the wire sticks through. Try this with different lengths of wire.

If solder is put in the wrong place, there are two ways to remove it. One is the use of a de-soldering pump; this is like a spring-loaded syringe in reverse. The plunger is pushed down and a clip holds it, the soldering iron is placed on the joint, and when the solder becomes molten, the de-soldering pump is placed at the side of the joint and the button holding the plunger down is pressed. This allows the plunger to return to its original position with a slurping action that sucks the solder away from the joint.

The second option is to use de-soldering braid. The braid is like a flattened stranded wire approximately 3 mm (0.125 in.) wide and is impregnated with flux. The braid is placed over the area where the solder is to be removed and the soldering iron is placed on the braid from above. The braid soaks up the solder like a sponge. This braid should be removed while still hot to prevent the braid sticking to the circuit board.

Holding Components in Place

Holding components in place while soldering has always posed a problem. One way around this problem is to purchase a "heat shunt." Despite its fancy name, a heat shunt is little more than a pair of tweezers in reverse. Squeeze a normal pair of tweezers and they close—squeeze a heat shunt and it will open. These are great for holding components in place while soldering. They have another equally important use. Some components such as transistors are sensitive to heat, so heat shunt tweezers are clipped onto the part of the transistor that is being soldered and the heat is "shunted" away. Small metal crocodile and bull clips can also be used to hold components in place (Fig. 5.2).

Holding the Circuit Board While Soldering

Another problem is holding the circuit board while soldering. When soldering, one pair of hands may not be enough. There's the soldering iron in one

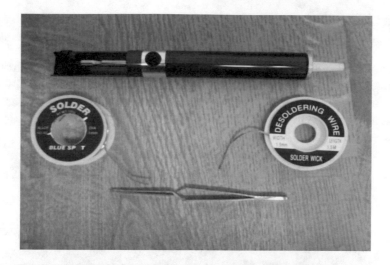

Fig. 5.2 Soldering equipment; Top centre de-soldering pump, right de-soldering wick, centre bottom "heat shunt" tweezers, left 1-mm diameter solder

hand, solder in the other, and then there's the components to hold, and on top of that, there's the circuit board as well. Four jobs, only two hands. The Fig. 5.3 is of a helping hand. This tool costs very little, and the magnifying glass is just an added bonus.

Fig. 5.3 A helping hand tool holding a piece of stripboard

Multimeters

Although not essential for the projects covered in this book, a multimeter can make life a whole lot easier when dealing with electronics, just by their sheer ability to find faults and the range of tests that they can perform. Multimeters come in two types: analogue and digital.

Fig. 5.4 Multimeters digital (left) and analogue (right)

Analogue Multimeters

These are usually cheaper to purchase, but it can take a little time to understand how to make a reading. The needle moves across the dial and there are a number of scales to read from. The trick is remembering to read from the right scale that matches the test being performed.

Most of the better quality analogue multimeters have a small mirror in the shape of an arc between two of the scales. The purpose of this mirror is to produce a reflection of the needle. When taking a reading, position oneself in such a way that the needle is covering its own reflection. This is then the correct reading.

Analogue multimeters can be (but not necessarily) less accurate than their digital counterparts. However, they are by far the better meter if there is a fault that results in a slow changing of values within the readings being taken, as the needle will respond to this, whereas the digital meter will just display a jumble of numbers.

Another downside to this type of meter is that they are less forgiving over polarity. If the leads are clipped onto a power supply the wrong way around, the needle rams itself against the stop. With a digital meter it will just show a negative sign at the side of the voltage reading.

Analogue meters are also less robust than digital meters. Handle them like a fine clock. The needle needs zeroing occasionally, especially after a knock. This is done by using a small screwdriver to turn an adjustment screw that is found somewhere on the body of the multimeter, usually near the point where the needle comes through the body of the meter itself. By making very small adjustments to this screw, the needle can be returned to its original position to read zero again.

Digital Multimeters

Digital meters can be more expensive to buy, but they are very easy to read. Some will even display a symbol for the test that is being carried out, such as A for amps when testing current, V for volts when testing voltage and Ω for Ohms when testing resistance.

These meters are more robust than their analogue counterparts (but still treat them with respect). They give highly accurate readings, good for when a precise reading is needed. They can be a bit of a nuisance with certain types of faults, as mentioned above, as they will just show a jumble of numbers on the display. They are also very forgiving, for example if a mistake in polarity is made.

Summary

A good multimeter can last a lifetime if properly looked after and can save lots of time and money around the home and car. It can be used for lots of things, from building radio receiver projects to sorting out the lights on a Christmas tree. Even if someone has just a passing interest in electronics, a multimeter is a must-have. As for the best, both have their merits, but just for the sheer ease of use, the answer must be a digital multimeter (Fig. 5.4).

An important point to note: if the multimeter is to be used to measure an unknown value, for example a voltage, start with the multimeter on its highest range and come down until a reading is found. Always read any instructions given with the multimeter before use.

Another good purchase to consider is a small project book explaining how to use a multimeter. These books can be bought from most electrical suppliers and cost very little. They give far more practical examples of how to connect a multimeter to an electrical circuit than most instructions provided with a multimeter. There are several on the market, an example being *How to Get the Most from Your Multimeter* by R.A. Penfold.

Electrical Component Identification (Fig. 5.5)

When speaking with others about radio astronomy, as soon as it is mentioned that one may need to build the receivers and that electrical components are involved, I have seen people's eyes widen and the color drain from their faces. At this point, their interest quickly evaporates.

Two of the projects covered in this book will require some self-building, namely the INSPIRE receiver and the Radio Jove receiver. If the art of sol-

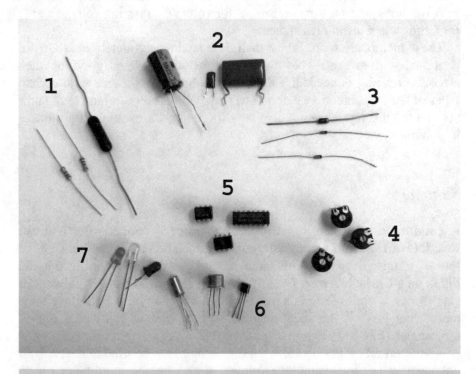

Fig. 5.5 Image showing a range of commonly used electrical components: (1) Resistors. (2) Capacitors. (3) Diodes. (4) Variable resistors and variable capacitors. (5) Integrated circuits (ICs). (6) Transistors. (7) Light emitting diodes (LEDs)

dering and component identification can be mastered, then building these two receivers will be no more difficult than building a plastic scale model of 100 pieces or more. No offence to Airfix and other providers of scale models, but the construction manuals supplied with the receivers are far better. Plus, once the receivers have been built, there is the added bonus of having something that works and can perform a useful task.

It is a nervous feeling after spending several hours building a project and the time comes to switch it on for the first time, but the nerves are quickly forgotten and replaced with feelings of excitement when everything works!

The only skill needed is to be able to identify each component, in many cases from a picture or image within the construction manual. The need to know how each component works to build the projects is not necessary, but it can be an advantage to know what job each component does. This can help with the installation of each part, as some can be fitted either way around, but others are polarity-sensitive and must be fitted in the correct orientation or they will not work. This is especially important in the case of electrolytic capacitors as they can burst open, covering the area in little bits of aluminum foil.

Many probably use these electronic components everyday without realising it. Take for instance the rotary volume control on a car radio—this is nothing more than a variable resistor. If someone is unfortunate enough to suffer a heart attack or other incident that stops the heart, the paddles placed on that person's chest to deliver the life-saving electric shock is a capacitor that has been charged up and then quickly discharged through the paddles into the body. The battery charger that is used to charge a car battery wouldn't work without the help of diodes that change the alternating current from the power supply into a direct current.

Here is a list of the most common components used in electronics and a brief description of what they do. In order to help anyone wishing to build either of the two receivers, the descriptions are accompanied by information on components identification, how to understand the orientation of components and whether the component is polarity-sensitive or not.

Resistors

The first component to be discussed is the resistor. Resistors come in many different sizes and shapes. There are two basic types: one of fixed value or one of variable value. They both do the same job, that is, to limit electrical current flow. Resistors are not polarity-sensitive and can be fitted either way around, but they have a power rating associated with them that is measured in Watts. The higher the wattage, the bigger the resistor will be.

The projects covered within this book use resistors of a power rating of 0.3 W and 0.5 W. The power rating should be taken into account if choosing to design electronic projects for oneself. For example, a favorite trick of the amateur optical astronomer is to use resistors to keep dampness from building up on the finderscope. This is done by soldering a number of resistors together and passing a current through them from a battery. As the resistors do their job and resist the flow of electrical current, heat is given off from the resistors themselves, and this heat is enough in some cases to keep the dampness away. All the resistors within the two kits are of the correct power rating.

To illustrate how a resistor works: take a garden hose with water coming out of the end. If the hose is then stood on, the water flow will be reduced. This would be equivalent to a resistor in an electrical circuit.

How to work out the resistance value of a resistor is needed. This is done by using a four-colored band code on the resistor itself. The Table 5.1 outlines how this system works.

The fourth band indicates the accuracy to which the resistor is manufactured, to the value shown.

Gold band = within 5%

Silver band = within 10%

No forth band within 20%

For example, let's we take a resistor with the four color bands of brown, red, orange and gold. Brown equals the number 1 in the first column and red equals the number 2 in the second column, so we have 12. If we then look at the fourth column to get the multiplier, we see that it is 1000. If we then multiply 12×1000, we get a value of 12,000 Ω. The fourth color band is the accuracy to which the resistor is manufactured, so a gold band would mean that the resistor is within plus or minus of 5% of the stated value.

Table 5.1 Resistor color coding

Color	First band	Second band	Multiply	Third band
Black	0	0	×	1
Brown	1	1	×	10
Red	2	2	×	100
Orange	3	3	×	1000
Yellow	4	4	×	10,000
Green	5	5	×	100,000
Blue	6	6	×	1000,000
Violet	7	7	×	10,000,000
Gray	8	8	×	100,000,000
White	9	9	×	---------------

It can take some time to get used to this system, but after building some projects, it will become almost second nature. If the colors of a resistor have become faded with age or if there is uncertainty regarding the value, a quick check with a multimeter is all that's needed.

Capacitors

Capacitors come in a wide range of types. Some are polarity-sensitive, some are not, and it is very important not to get this wrong. Capacitors are usually marked in some way if they are polarity-sensitive. They can be of fixed value or variable type, like a resistor. Unfortunately, unlike resistors, there is no universal color coding. Much depends on the manufacturer, and while some use a form of color coding, other manufacturers stamp the value somewhere on the capacitor itself. The maximum working voltage the capacitor can withstand should also be marked on the capacitor. The value of the fixed capacitor will always be the same, even if the voltage that it is used at is lower than the maximum working voltage. Never exceed this maximum working voltage.

The job of the capacitor is to act like a reservoir, but instead of storing water, capacitors store electrons. They can be used in timer circuits in conjunction with a variable resistor, where the variable resistor is used to slow the charging of the capacitor before triggering other actions to take place, such as the intermittent control on a car's windshield wipers. Capacitors can also be used as filters to smooth out spikes within an electrical circuit. Spikes are absorbed within the capacitor itself, in a similar sort of way that the flywheel on a car's engine smoothes out the motion of the pistons to keep the rotation smooth and constant. It is important to remove these spikes, especially with radio receivers, in order not to artificially introduce unwanted noise within the receiver's electronics.

Variable capacitors are used in the tuning circuits for radio receivers like the ones covered in this book. They are set by using a small screwdriver, which turns movable plates to alter the size of the storage area. One of these variable capacitors will need to be set after building the Radio Jove receiver.

Capacitors must be respected at all times, as they can still hold an electrical charge for some time after the power has been switched off. The old cathode ray tube televisions are particularly bad for this, as they could hold a charge for days if not longer. Never poke about inside any electrical appliance with anything, least of all a metal screwdriver, as doing so may necessitate the help of a capacitor to restart one's heart! The really dangerous capacitors are the high voltage electrolytic type. An anecdotal warning tale:

the author once witnessed a foolish act by an apprentice electrician, who short-circuited one of these high voltage electrolytic capacitors to the side of his metal work bench. The capacitor in question was as big as a beer can. There was a bright flash and the capacitor's terminals welded themselves to the side of his work bench. He had to hacksaw through the capacitor terminals to remove it. The workshop foreman, as can be imagined, wasn't very happy, and neither was the apprentice when he was charged for a replacement capacitor.

The capacitors used in the two projects covered within the book are of low voltage, so no harm should come to those who carefully follow the instructions given within the construction manual provided.

Diodes

Diodes come in many different types, but in its simplest form, a diode's job is to act as a non-return valve in an electrical circuit. Diodes allow electrons to flow one way, and they will block any movement in the opposite direction. It is therefore important that diodes are fitted in the correct orientation. Looking at the casing of a small signal diode, it will be noticed that there is a color band at one end. The band denotes the direction of current flow. Remember that the current flows through the diode towards the band. Then the correct orientation of the diode will be known.

Diodes come in different power ratings, like resistors, and this power rating must not be exceeded. Looking at a few of the diodes available will give a better understanding of the different uses of each type.

Power-rectifying diodes are used to change alternating current into direct current, and also to help control power spikes in high-powered switching applications. Depending on the power and voltages that they are being used for, it is not uncommon to find such diodes surrounded with substantial metal heat sinks to dissipate any heat they produce and keep them cool.

Small-signal diodes are the sort of diodes that will be used within the projects described in later chapters. They are the little brother of the power-rectifying diodes, and while they perform the same sort of job as their big brother, they are much smaller and can be mounted on a circuit board. Diodes can be used singly or in conjunction with other diodes. If one diode is used to rectify an alternating current into a direct current, the diode blocks the reverse current flow, so only half of the alternating current gets through. This is known as *half-wave rectification*. Four diodes can be used in conjunction with each other to form what is known as a bridge rectifier. This has the advantage of allowing all of the alternating current through, but in such a way that it has now all been changed to direct current. This is known as *full-wave rectification*.

Zener diodes are a special type of diode. Unlike the other diodes discussed above, which will block all reverse current flow, zener diodes will, under certain circumstances, allow a reversal of current flow to take place. Different voltage values can by assigned to the zener diode reversal quality. This means they can be used as a voltage regulator or a voltage-sensitive switch to protect other components from overloading.

Light-emitting diodes, or LEDs for short, are another special type of diode, but they are made of a different substance to the other diodes described above. This material emits light when an electrical current is passed through it. As with all other diodes, they are polarity-sensitive and must be fitted in the correct orientation. Looking at an LEDs plastic cap, you will see a rim at its base. Upon closer inspection of this rim, you will note that there is a flat mark on the rim itself. This denotes the negative connection.

LEDs come in a number of different colors and are used in just about every electrical device made, from the standby light on a television set to the International Space Station.

There are several advantages of LEDs over normal incandescent lightbulbs. LEDs have a much longer life than a normal lightbulb, in some cases 10,000 times as long, and they use far less power than a normal light bulb. This is why lights on a car are slowly being changed to LEDs, and now that the white LED has been developed, it can only be imagined what else will be replaced with LEDs. An LED is far more robust than a normal lightbulb, and unlike a normal lightbulb, an LED doesn't mind being switched on and off repeatedly. LEDs also operate at a far cooler temperature, so unlike a normal lightbulb, there are no burnt fingers when handling them.

Transistors

Transistors come in all different shapes and designs, from small plastic ones right up to the hefty metal-bodied power transistors. In fact, there are whole books listing hundreds, if not thousands, of transistors and some two dozen different uses for them, such as amplifiers, high frequency switching, and power switching, just to name a few.

Each transistor is designed to have its own special quality and use. Some are small in size, such as the small signal and switching type used in the construction of the two receivers covered within this book. Others, like power transistors used in high power amplifiers, are quite large and need to be fixed to a heat sink to keep them cool.

Transistors have three wire connections coming from them: a base, collector and emitter. Some transistor designs have only two wire connections,

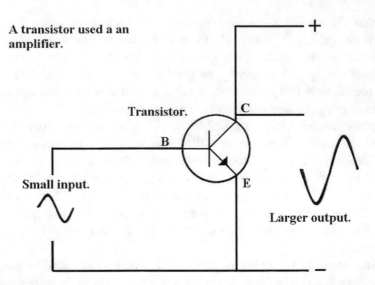

A transistor used a an amplifier.

Transistor.

B

Small input.

C

E

Larger output.

+

−

Fig. 5.6 A transistor used as an amplifier

like the high power transistors, and these use the body of the transistor itself as the third connection. The basic principle of operation entails a small current or voltage applied to the base connection, allowing a larger current or voltage to be passed through the collector and emitter connections. See Fig. 5.6, which shows the connections base, collector and emitter as B, C and E, respectively (other components, such as resistors, have been left out for clarity).

This means that a transistor can be used as an amplifier. If a small input signal, such as the weak signal picked up from an antenna, is fed into the base of the transistor, this will cause a large current to flow through the collector and emitter. If this process is repeated through different stages, the input signal will be amplified to a level that can eventually be used to drive a speaker.

Most of the small signal and switching transistors are quite delicate and can easily be damaged if they are overloaded with too much voltage or current. It goes without saying that transistors must be fitted in the correct orientation. Some transistors have the connections marked on the body of the transistor itself, but others need data sheets to sort out the connections. The transistors used in the two receivers covered in this book come with very clear instructions and it is unlikely there will be any trouble fitting them in the correct way. Another important point to be taken into consideration is that transistors can be damaged by heat, so it is important when soldering them in place not to leave the soldering iron in contact for longer than necessary to solder the connection. It is also a good idea to use·a heat shunt to protect the transistor from any unnecessary heating during the soldering process.

Inductors

An inductor, sometimes called a choke, is in its simplest form a coil. It can take on a number of different designs, from some that look similar to resistors to wire-wound coils. The physical size and shape of an inductor takes into consideration two important factors: the power rating and the level of inductance required. The inductors used with the two projects covered here have the appearance of a capacitor, but are a dark grey in color, and they are clearly labeled and not polarity-sensitive. Without going into too much detail about how inductors work: as an electric current flows through the coil of the inductor, a magnetic field is generated. The effect of the magnetic field produced by the current flow is called *inductance*. Inductors are used in conjunction with capacitors to form tuning circuits, for example the tuning circuit within a radio receiver, or as a filter that will only allow the desired frequencies to enter a circuit.

Integrated Circuits

Integrated circuits, ICs for short, are sometimes called silicon chips or microprocessors. These are the small, black, rectangular-shaped items inside just about every electrical appliance on the planet.

Instead of having lots of different components such as transistors, resistors and diodes soldered separately onto a circuit board, an IC has all these components etched on to a single wafer of silicon, hence the name silicon chip. Each wafer of silicon can contain anything from a handful of components to many thousands. The invention of the silicon chip made it possible to downsize items, such as a computer, from a bulk that filled an entire room to something as small as or smaller than a briefcase. The ability of having thousands of components on one small wafer of silicon has also increased computing power vastly over the years since its creation. In the early to mid-1980s, computers such as the Sinclair spectrum ZX 81 were on sale boasting of their 1 K of memory, but now it is not uncommon for a cellphone to contain a 64 Gb memory card that is half the size of a postage stamp. To put this in perspective, when the Apollo 11 crew went to the

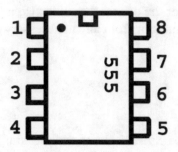

Fig. 5.7 Image of a 555 IC used to show orientation

Moon in July 1969, the computing power they had at their disposal was infinitely less than the computing power of the cellphone we now keep in our pockets.

Integrated circuits are supplied in many different types, and they can be used to build timer circuits, amplifiers and just about anything that can be imagined. To build these projects, we must know how to orientate an IC and place it on the circuit board the right way up. All ICs, whatever the manufacturer, have the connections numbered and are orientated the same way. Looking at Fig. 5.7, we see an eight connection IC known as a "555" that is used in conjunction with other external parts such as resistors and capacitors to produce timer circuits. The internal workings of the IC are not important here.

Take note of the small notch at the centre. This notch indicates the top of the IC. To the left of this notch a small dot will be seen. This mark can be a white spot or even a small indentation. However it is marked, there should be a mark there. This mark shows the position of the number one connection. They are then numbered in the same way as on the image. If the IC has more than eight connections, for example 16, then the number one connection will be at top left, then numbered down to the eighth connection bottom left, and then across to bottom right (this will be number 9), back up to the top right to number 16.

Integrated circuits can be very delicate, and some can even be damaged by the static electricity contained within the human body. An IC of this type should come in a small pack bearing a static electric warning on it. In such circumstances, an earthing strap should be worn around the wrist with the other end connected to a suitable earth connection before handling the IC. Thankfully none of these types of ICs will be used here.

ICs can be very easily damaged by the heat from soldering, even more so than transistors. One way that is used to stop this heat damage is not to solder them. What is used instead is an IC socket, a socket with the right number of connections on it. This socket is then soldered onto the circuit board in place of the IC itself. The IC is then plugged into the socket. This has another advantage that if the IC itself goes wrong for whatever reason, it's just a matter of unplugging it to replace it with a new IC.

There is a very important point to make about plugging in and unplugging an IC from its socket. IC connections are very fragile and can easily be bent or broken, so be very careful when plugging an IC into its socket to make sure that the connections on the IC go into the correct corresponding connections on the socket. This may seem to be stating the obvious, but the connections on an IC are very close together, and mistakes can easily happen.

Sometimes the connections on an IC are spread open too much to fit it in to the socket. If this is the case, carefully bend in the connections of the IC so it can be fitted into the socket. A good way to do this is to hold the IC between the finger and the thumb and very carefully push the connections down on top of a table. This will bend all the connections equally, and hopefully keep them all in line. Something that can sometimes turn into a bit of a nightmare is the unplugging of an IC from its socket, especially if it has been in the socket for some time. Luckily this happens only on very rare occasions. It's not a simple matter to pull the IC out—it must be teased out of its socket. There are special tools that can be purchased to do this job. Another good way is to use two very small flat-bladed screwdrivers to fit under each end of the IC, then slowly ease up each end a little at a time, in a sort of rocking motion. Don't be too enthusiastic with either screwdriver as the IC connections can bend or even break, or even worse, the IC casing could split open.

Treat an IC as if handling a stick of dynamite, and then nothing should go wrong. To save the risk of you accidentally damaging an IC, they should be one of the last components fitted when building up a circuit.

Headphones

Headphones will be useful for the projects covered within this book. They can sometimes be overlooked, but the value of a good pair can't be overstated because they can make listening a much more pleasurable experience. These need not be expensive. By shopping around it is usually easy to pick up a good pair quite reasonably. Try and avoid the fiddly little ones that come with MP3 players and the like, as some of these are better than others. Avoid the really cheap headphones, as their sound quality is poor and any noises will sound like they are being emitted inside a tin can. Headphones that are labeled as having a good bass are better, and they are made from a wide range of materials, from plastic to silicon. These can be used with the projects within this book if nothing else is available, but they are really only any good for the use for which they are intended—something lightweight that fits in the ears for 30 min whilst out jogging in the park.

Full-sized headphones are better and are generally more comfortable to wear. Try to get a pair with a cushion that fits all the way around the ear, as opposed to those that just sit on the ear. This has two good advantages: it helps keep outside noise out and keeps the noise from the speaker inside. It gives the effect of being in a soundproof room. A good way to test this is to try and listen to some music in a room where the television or radio is switched on using the little earbuds. Without changing the volume setting, swap to the full-sized headphones with the cushion that fits around the ear and enter the same room. The sound from the headphones will be clearer and the noise from the television or radio will be less of a distraction.

All the sounds that will be heard from the projects covered in this book will be heard against the constant background hiss of white noise, and the advantage of a good set of headphones can't be stressed enough, as it can make all the difference for hearing a sound or not hearing it.

Unlike listening to music, where just looking at the CD case or device screen will tell you what the next track will be, the radio astronomer doesn't have this luxury when sitting around listening for a meteor ping, as they come completely randomly and can happen any time.

Here are other things to look for in a good set of headphones:

- **The driver unit**, which is basically the size of the speaker in each earpiece. This should ideally be something along the lines of 32 mm (1.25 in.) to 40 mm (1.57 in.).
- **Frequency response**. All headphones have something along the lines of 20 Hz to 20 KHz, as this corresponds to the audio frequency of human hearing.
- **An inline volume control** is handy, but not essential, as it allows the volume to be increased or decreased without touching the receiver. This is handy if the receiver has been calibrated to the computer, as any changes to the receiver volume will mean the calibration procedure will have to be done again.
- **The length of cable** that comes with the headphones may seem unimportant, but it is crucial. One metre (39 in.) is not long enough. The cable will be at full stretch, and the slightest movement could pull the headphones off the head, or worse, pull the receiver off the desk and send it crashing to the floor. Three metres (117 in.) is too long, as it would likely

tie itself in knots and get wrapped around anything and everything. Two metres (78 in.) is just about right, as this length would allow the user to move about without getting the cable wrapped around everything.

Full-sized headphones usually come with a 3.5 mm (0.125 in.) or 6.3 mm (0.25 in.) stereo jack plug fitting. The most common fitting used on MP3 players, cell phones, computers, etc. is the 3.5 mm (0.125 in.) jack plug. All the projects covered here require the smaller of the two plugs. Headphones sometimes come with an adaptor that can be used to change from one size to the other. The adaptor just plugs on the end, either to make the connection to the item larger or smaller. If an adaptor is not included with the headphones, one is easily obtainable from most electronics shops or online outlets for a nominal outlay.

The final, and probably the most important thing to remember, is to make sure your headphones are comfortable. This is what makes in-person retail shopping advantageous. Before purchasing a pair, ask to try them on. If they are too heavy, look around for a lighter pair. This is important as they may be worn for many hours at a time and you will become sore. The weight is sometimes stated on the box.

There are new wireless headphones on the market nowadays. These work by having a small transmitter, such as Bluetooth, built into a device, or a transmitter that plugs into the headphone socket to transmit the music. This transmitter turns the audio signal into a radio frequency, and then a receiver built into the headphones turns the radio frequency back into an audio signal, which is then played through the earpieces of the headphones, thus eliminating the need for a cable. This may seem ideal, but the thought of having a radio transmitter beaming a radio frequency signal to the headphones in close proximity to a sensitive radio receiver doesn't seem to be a great idea.

As interference is the greatest enemy of radio astronomy, it seems sensible not to introduce any unnecessary interference either from the signal, from the transmitting unit or from the electronic circuitry contained within the transmitter unit itself. Another disadvantage of this type of wireless headphone is that it needs batteries to power it. Also, batteries have a habit of running out at the most inconvenient times.

Noise-canceling headphones work by electronically canceling out sources of outside interference. Reviews of some of these headphones have been good, but these reviews are based on listening to music or being on an airplane, rather than listening to weak radio signals from space.

A well-chosen pair of headphones need not be expensive and can be used for anything, not just radio astronomy. If looked after properly, they can last a lifetime (Fig. 5.8).

Fig. 5.8 This image shows a full-sized pair of lightweight headphones with good soundproofing cushions that fit all the way around the ears. There is also an inline volume control. This pair was not expensive to purchase. They are made mostly of plastic, but have soft leather ear cushions and deliver excellent sound quality

Chapter 6

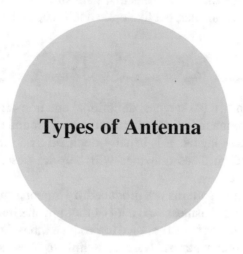

Types of Antenna

Antennas

An optical telescope uses either a mirror or a lens to collect light. The light is then brought to a focus and magnified by an eyepiece. An antenna in some respects can be likened to an optical telescopes lens or mirror, as the antenna collects the radio signals and brings them to a focus at the dipole of the antenna. This signal is then passed along the feed cable to a receiver and amplified.

Optical telescopes come in three main types: a *reflector*, which uses a mirror to collect light, a *refractor*, which uses a lens to collect light, and a hybrid, such as a *Cassegrain*, which uses a combination of a lens and a mirror to collect light.

All telescopes can be used for both deep-sky and planetary observing, but each telescope is better at viewing some objects than others. For instance, a planetary observer may choose a refractor over a reflector. A refractor will have less light-gathering capabilities—not a problem with bright objects like planets, but a drawback with deep-sky objects. On the plus side, the observer will be rewarded with high-contrast views of planets.

On the other hand, a deep-sky observer may opt for a reflector over a refractor. A reflector, having a large mirror, will collect more light than a refractor, therefore allowing an observer to see fainter objects. The down side is that the contrast may not be as good as a refractor.

S. Arnold, *Radio and Radar Astronomy Projects for Beginners*, The Patrick Moore Practical Astronomy Series, https://doi.org/10.1007/978-3-030-54906-0_6

The secret to observing is knowing which telescope will work best for the job you wish to perform. The same is true with antennas; some will work better for one task over another, but like optical telescopes, they can overlap.

Antenna Radiation Patterns

Antennas come in two basic types: directional and non-directional (omnidirectional). An antenna radiation pattern is the area around the antenna where it picks up the most signal. It is usually depicted as a drawing that shows lobes. Understanding these drawings will help an observer get the most from their antenna.

Antenna radiation patterns are produced by antenna modeling. Antenna modeling is a serious business, and a lot of thought and mathematics go into its design. Software has been designed to produce three-dimensional images of antenna radiation patterns. Here is a link to download free antenna modeling software to experiment with: www.qsl.net/4nec2/. Once there, download 4nec2 file.

After being installed on a computer, the software will generate a three-dimensional image in the X, Y and Z axis. This image can then be rotated around. This software is not for the absolute beginner, as values for the antenna need to be known. A four-part tutorial for this software can found at the following links:

- Part 1: http://www.youtube.com/watch?v=6T58CLyqQa0
- Part 2: http://www.youtube.com/watch?v=3N_IbG1-InA
- Part 3: http://www.youtube.com/watch?v=x7HaLvQARMQ
- Part 4: http://www.youtube.com/watch?v=wca6_5g7sQe

Here we are only going to touch on the basics, and the mathematics will be kept to a minimum. In this way, we can use an antenna pattern to our advantage. For example, let's say an antenna is picking up interference and performing badly. If we know in which direction the source of the interference is, and know where the minimum pickup of the antenna is, by pointing the minimum pickup area of the antenna towards the source of the interference, we can remove some of its effects and allow the antenna to perform better.

Half-Wave Dipole Antenna

A *half-wave dipole antenna* is so called because the length of the antenna is equal to half the wavelength of the frequency at which it is to be used. The feed cable will be in the centre. This is the type of antenna used to pick up the Sun and Jupiter.

Let's say we have a dipole antenna in a horizontal position hung in free space. By free space, we mean the antenna is hung at least one wavelength above the ground and strung between two poles. In this case, the antenna pattern will look like what is shown in Fig. 6.1. The two lobes represent the antenna radiation pattern. To picture this in three dimensions, think of a large bagel or an inflated inner tube. This pattern is called a *torus*. The antenna is the horizontal line in the centre.

The maximum pickup or gain of the antenna is at the top and bottom of the lobes, 90° to the antenna. The minimum pickup or gain is in line with the antenna, or left and right in the image. If an antenna can be positioned in such a way that the minimum gain part is pointed at a source of interference, it would improve the overall performance of the antenna.

If the antenna is used vertically, as long as the height of the antenna is equal to one wavelength, the radiation pattern would be the same but turned through 90 degrees.

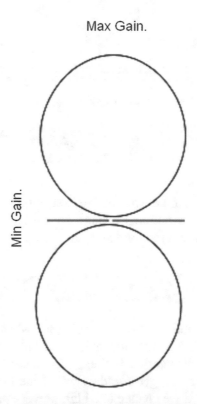

Fig. 6.1 Radiation pattern of a single dipole antenna

The Whip Antenna

This type of antenna is used on cars and portable radios. It can also be used to pick up communications with the International Space Station (ISS). It has a similar radiation pattern as the half-wave dipole, but with a twist, as now we have the ground effect to consider. The ground effect will become important later when discussing the twin dipole used to pick up The Sun and Jupiter. The ground effect changes the shape of the radiation pattern of an antenna. Let's use the example of an inflated inner tube, but this time lay it horizontally on a hard surface such as a table. If we then push down on the inner tube so the side on the table is flat, the displaced air inside the inner tube has to go somewhere, so it pushes the lobes up higher into a dome shape. See Fig. 6.2.

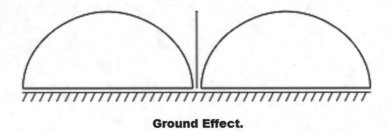

Ground Effect.

Fig. 6.2 Radiation pattern of a ¼ wavelength vertical whip antenna showing the ground effect

This ground effect can be exploited, as varying the height of an antenna can make the radiation patterns change so we can make the maximum pickup area point in the direction we want.

The Yagi Antenna

A Yagi antenna has multi-elements, ranging anywhere from two upwards. The more elements it has, the higher its gain but the narrower the beam width. Using an optical example, if a telescope is used to view a planet with a low-power eyepiece, we get a wide view of the sky but the planet will show as a small disc. If we change to a high-power eyepiece, we get a good size planetary disc but not much sky. Therefore it's a trade off between magnification and field of view.

Yagi antennas are directional antennas, unlike the ones already discussed. They are used for receiving television pictures. We will be using a three-element Yagi to pick up meteor echoes as well as echoes from the ISS, and a ten-element Yagi to pick up echoes from the aurora.

Let's first explore a *three-element Yagi*. The first element is the detector, sometimes called a parasitic element. Its job is to detect incoming radio waves and steer them onto the third element. The second element is the dipole where the feed cable is connected. The third element is the reflector, which focuses any radio waves back to the second element the dipole.

Let's take a look at the workings of a Schmidt Cassegrain (SCT) optical telescope. The detector element would be the SCT's corrector plate, steering light to the primary mirror—in our case the reflector, the third element. This light is then reflected to the secondary mirror—our dipole, the second element. From the secondary mirror in the SCT, the light is brought to a focus and magnified by the eyepiece—in our case a radio receiver.

The properties of the three-element Yagi antenna make it highly suitable for picking up meteor echoes. It has more gain than a single dipole but still has a good beam width. Looking at Fig. 6.3, the second element will be where the two ellipses meet, while the third element would be behind the small ellipse. The first element would be pointing towards the maximum gain.

To visualize the radiation pattern of a three-element Yagi in three dimensions, let's imagine inflating a tear-shaped balloon and holding the balloon at the bottom where it has been tied. The radio waves would be traveling towards the person holding the balloon and passing through the top of the balloon towards the base.

3 element Yagi.

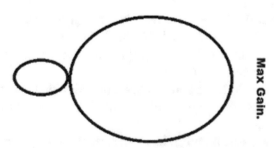

Fig. 6.3 Radiation pattern of a three-element Yagi antenna

Fig. 6.4 Radiation pattern of a ten-element Yagi antenna

Now let's compare the radiation pattern with the ten-element Yagi antenna used to pick up echoes from the aurora. Figure 6.4 shows the radiation pattern of a ten-element Yagi antenna. The reflector element and the dipole element are in the same place, but instead of having one detector element, the ten-element Yagi has eight detector elements. This has the effect of narrowing the beam width but increasing the gain in one direction.

In Fig. 6.4, this is shown by the lobe being thinner in width but longer in length.

To imagine this radiation pattern in three dimensions, think of a balloon 100–125 mm (4–5 in.) in diameter and about 600–760 mm (24–30 in.) in length. If you held the balloon by the tie at the bottom, the radio waves would travel from the top of the balloon down its length to where it is tied.

Like dipole and whip antennas, Yagi antennas are impacted by ground effects, which can change the shape of the radiation pattern, but at the wavelengths used in this book these can be ignored.

Antenna Gain

There are several methods to express the gain of an antenna. Here we are going to use the *dbi method*. The "db" stands for "decibel," a numerical factor to represent the antenna's gain. The "i" stands for "isotropic," or the gain of a theoretically perfect antenna.

An isotropic transmitting antenna has a radiation pattern that looks like a sphere and radiates equally in all directions. A good way to imagine this is to think of the Sun.

The Sun radiates heat and light in all directions—up, down, left, right, forwards and backwards. Therefore the Sun could be considered as isotropic transmitter of heat and light. An isotropic receiving antenna is the opposite. The radiation pattern is still a sphere, but unlike the isotropic transmitting antenna, which transmits equally in all directions, the isotropic receiving antenna receives signals equally in all directions. If the Earth were an isotropic receiver, its surface would have equal temperature and daylight all the time everywhere on Earth.

The gain of a real antenna is compared to that of the theoretical isotropic antenna. Looking at the radiation pattern of the single dipole antenna above, it can be seen that it is not a sphere and has a gain in one direction over the other, which is shown by the minimum and maximum pickup. The gain of the single dipole is the difference between the maximum pickup area and the theoretical isotropic antenna, around 2.1 dBi.

If we now look at a Yagi antenna, which is a directional antenna, the gain can be more easily seen. It has already been said that the more detector elements, the more directional and the higher the gain, but the beam width is reduced.

Looking at Fig. 6.5, it can be seen how the beam width of an antenna reduces as the gain increases.

Gain is measured on a logarithmic scale, therefore every increase of 3 dBi doubles the gain of the antenna. For example, 6 dBi is double the gain of 3 dBi and 9 dBi has double the gain of 6 dBi, and so on.

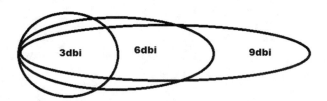

Fig. 6.5 Image showing the change in beam width of an antenna as the gain increases

Chapter 7

Cables and Connections

Feed Cables and Connections

The job of the feed cable is to carry the signal from the antenna to the receiver. The feed cable is made up of an inner conductor, which can be a single solid wire or a multi-stranded one. Surrounding the inner core there is the inner dielectric insulator, then a woven copper shield and finally an outer plastic covering.

There are a number of coaxial cables on the market. Here is a link to a downloadable catalogue of coaxial cables along with their specifications: http://www.helukabel.com/CATALOGUES/kap_M_336-347_en_link.pdf.

Each cable has its own specifications. Some have less loss of signal at certain frequencies than others. If building a project, the best thing to do is pick the cable that suits the job at hand.

The original version of this chapter was revised. The correction to this chapter is available at https://doi.org/10.1007/978-3-030-54906-0_20

S. Arnold, *Radio and Radar Astronomy Projects for Beginners*, The Patrick Moore Practical Astronomy Series, https://doi.org/10.1007/978-3-030-54906-0_7

Feed Cables: Does Cable Length Really Matter?

If there is one thing that puts the cat amongst the pigeons with amateur radio enthusiasts, it is the question of what length feed cables should be. Some regular radio users reading this will feel strongly about the subject and will fall into one of two groups.

Group 1 thinks it doesn't matter what length the feed cable is, as it only carries the signal to the receiver.

Group 2 thinks the feed cable must be cut in multiples of a full or half a wavelength, from a minimum of half a wavelength up to a maximum of five wavelengths of the frequency that is to be received. With the projects described in this book, five times the wavelength is used as a maximum length for a feed cable; any benefit beyond this will be lost due to losses within the cable itself. This will change at higher frequencies and consequently at shorter wavelengths being received or if amplifiers are being used.

For example, let's say a frequency had a wavelength of 10 m. This means the feed cable would need to be cut at multiples of 5 m, from a minimum of 5 m up to a maximum of 50 m. That is if the signal traveled down the cable at the speed of light with no resistance—which it doesn't—so we must take into consideration the velocity factor of the cable being used. All coax cables have a *velocity value* to them. Basically, this is the speed at which the signal travels down the cable. A velocity factor of 0.8 means the signal travels at 0.8 times the speed of light down the cable. This needs to be known to work out the true length of cable to use. So in the above example where a wavelength of 10 m was used, if we factor in the cable velocity of 0.8, this would change the length of the cable. In this case, the cable would have a minimum length of 4 m and a maximum length of 40 m.

Let's do a practical example of this from scratch, say, if we were to cut a feed cable for a frequency of 144 MHz. This can be easily worked out in the following way:

$$\frac{\text{Speed of light}\,(\text{kilometres per second})}{\text{Frequency}\,(\text{KHz})} = \text{Wavelength}\,(\text{metres})$$

The frequency must be changed to KHz, so if we are working at a frequency of 144 MHz (multiply 144 by 1000 to turn it into 144,000 KHz):

$$\frac{300,000\,(\text{kilometres per second})}{144000\,(\text{KHz})} = 2.083\,\text{wavelength}\,(\text{metres})$$

If the coax cable being used has a velocity factor of 0.8, you would take the wavelength and multiply this by the cable velocity factor to determine what multiples to cut the cable into:

$$2.083 \times 0.8 = 1.6664 \, \text{m} \left(\text{round this up to} \, 1.67 \, \text{m} \right)$$

Therefore the cable should be cut in multiples of this figure, from the minimum of half a wavelength, 0.835 m up to a maximum of five wavelengths, 8.35 m.

Both groups agree on one thing—that the other group is wrong. So who is right and does it matter?

Theoretically, if cables transmitted 100% of the signal without loss, it shouldn't matter how long the feed cable is. Well in theory, the Titanic was unsinkable, and that didn't end very well for anyone. Hopefully this explanation should help settle the argument (or make it worse).

First let's look at an everyday practical example—take fitting a car radio. Will the length of the feed cable really make that much difference? Most radio stations are *frequency modulated* (FM). A radio is designed to pick up more than one station. Each radio station broadcasts on a different frequency. The radio needs to be capable of being tuned to different frequencies. In this case, when installing a car radio designed to receive a number of powerful modulated radio signals, the argument of feed cable length becomes somewhat irrelevant, as any loss of signal due to cable length will be so minimal it will go unnoticed.

With radio astronomy, the received signal is unmodulated and naturally varies widely in strength, but is centered around one frequency. Cutting a feed cable to the correct length is effectively "tuning" the cable to that single frequency. Experiment and experience have shown it is worth taking a little extra time to work out the correct cable length.

Here is a true story to demonstrate this. A few years ago I was observing the August Perseid meteor shower with a friend. The equipment and frequency being used meant the feed cable length needed to be 6.8 m (22.3 ft) in length. We were receiving good, strong signals with persistent echoes. It was suggested that using the shortest length of feed cable possible may make the echoes even stronger, as the signal had less cable to travel along. This argument seemed sound at the time. In a mad scramble, the spares cupboard was raided and a much shorter cable was made. The longer cable was changed for the shorter one, and the screen was watched with anticipation of receiving stronger louder echoes. Surprisingly, with the shorter feed cable there was a weaker, inferior signal and the meteor echoes had all but gone. The cable was removed and checked for defects; it was fine. The longer cable was refitted and the signal strength came back!

One can argue that there may have been fewer meteors at the time of using the shorter feed cable, and that it was just a coincidence that when the longer cable was refitted, the signal strength returned. So, the challenge is to try this simple experiment yourself! It would be interesting to see if the results are the same and even more interesting if they are not.

It helps knowing values for cables and other equipment, as this can be useful later if calculations are to be made on signal strength and if the project is to be taken to the next level of advancement. The signals that will be received in the projects described later will be a lot weaker than a local radio station, so any help no matter how small will be a good thing. A good rule to follow is: 1% of something is better that 100% of nothing.

Types of Cable Connections

When building antennas, a good idea is to choose one type of connection and stick with it, as this will prevent the need for countless adaptors. At the frequencies being used here, a BNC and F-type connector will be adequate. Don't be tempted to use PAL (Belling Lee) connections commonly used on televisions. These connections start to break down at around 150 MHz and get very noisy.

BNC Connection

BNC connections cover a frequency range from 0 to 4 GHz. These are good, solid connections. They fit together with a bayonet-type fitting and can quickly be coupled and uncoupled, which can be handy when working at night.

BNC connections can be either crimp-on, solderable or twist-on. If using crimp-on connections, a crimping tool will be needed. Solderable BNC connections are popular with radio hams but need a soldering iron to fit them. Twist-on BNC connections are sometimes frowned on by radio hams, but if fitted correctly, and if the connector is not subjected to a lot of weight from cables, etc., they are surprisingly hardy. BNC connections are good if the central conductor is a stranded one.

Here is links showing how to fit the different types of BNC connector:

- Crimp-on BNC plug: https://www.youtube.com/watch?v=ktQVwfo-s9w
- Twist-on BNC plug: http://www.youtube.com/watch?v=BNO81-RU5ZY
- Solderable BNC plug: http://www.youtube.com/watch?v=DksUM656s-m

F-Type Connection

F-type connectors cover frequency ranges from 0 to 1 GHz. F-type connections are the screw-on type. They are good if the central conductor in the coax cable is the solid type. They have a couple of drawbacks:
1. The central connection of the coax cable is exposed, so care needs to be taken coupling and uncoupling these types of connections so as not to damage the central conductor.
2. As mentioned above, F-type connections are screw-on type, therefore they can be a little fiddly when using them at night.

Here is a link showing how to fit an F-type connector: https://www.youtube.com/watch?v=Bn30LdXiWx4.

SDR Receiver Connections

After choosing which connection to use, a corresponding adaptor is needed to couple the chosen connection to the SDR receiver. These can be purchased easily online for a few dollars.

There are a number of SDR connections. SMA (Sub Miniature version A) connections are the best. These are a screw-on type. They are quite strong and once fitted can be left on the receiver and different antennas fitted as and when needed. A word of caution about SDR receivers that have a MCX (micro-coaxial) connection: this type of connection is small and can be a little fiddly. The adaptor takes quite a bit of force to fit it into the dongle; a loud "click" is heard as it's fitted. Rather than getting a "straight through" adaptor, a better chose is a "jumper" adaptor. A jumper is a short length of cable with the connections on either end of the cable. These cost a few dollars more but are worth the expense. A tip is to tape the jumper cable to the SDR dongle; this has the advantage of taking the weight off the MCX connection.

Chapter 8

Shielding of Equipment

Shielding of Cables and Ferrite Collars

There are a few things that can be done to reduce or stop interference. Firstly, don't use old cables with cracked or damaged outer coverings, as moisture and water can leak into the cables. Don't lay cables in areas of known sources of interference. This may seem obvious but it is surprising how many times it happens.

A friend once commented that his television kept crackling every time he had his central heating on. On examination of his system, it was found that he had inadvertently run his antenna cable too close to his boiler, and every time the spark ignited the gas in the boiler, the television crackled. Rerouting the cable removed the problem.

There is a small, passive item that can be added to cables to help prevent some of the problems associated with interference—these are called *ferrite collars*. Look at the end of a cell phone charger cable or the end of a computer power cable. The big cylindrical-shaped object on the end is the ferrite collar. The ferrite collar's job is to reduce electrical interference. It does this by absorbing any electrical interference within the cable that is passing through the collar and stopping any electrical interference entering the cable at its weakest point (the open ends). They release the absorbed energy in the form of heat and may feel warm to the touch.

S. Arnold, *Radio and Radar Astronomy Projects for Beginners*, The Patrick Moore Practical Astronomy Series, https://doi.org/10.1007/978-3-030-54906-0_8

Ferrite collars can be used at both ends of coax feed cables, low-voltage power cables and audio cables, as a preventative measure to stop problems before they start. Ferrite collars come in many different sizes. If they are being used on a coax cable, it is better if the collar size is as close to the size of the cable as possible. Measure the outer diameter of the cable on which they are going to be used to get the closest matching size. If they are slightly loose when they are fitted to the coax cable, simply apply some PVC electrical tape to the cable to take up the slack. A good idea is to fit ferrite collars on both ends of coax cables, as the ends are the weakest point and where interference is likely to enter the cable.

When using ferrite collars on a power cable (*not* MAINS power cables), it's perfectly fine just to loop through the ferrite collar and pass the power cable back through itself so there is no need to cut off the end, or to otherwise get a clip-on ferrite collar. These open up and let the cable pass through the middle, then clip shut. These are a little more expensive but are ideal if you do not want to remove the ends of the cable. On a power cable you only need to fit a ferrite collar to the end that goes into the apparatus, similar to a computer power cable.

Audio cables should have a ferrite collar fitted to each end (see Fig. 8.1). They can be bought online, or keep an eye open at radio shows for old cables with ferrite collars on the ends. These can be picked up quite cheaply.

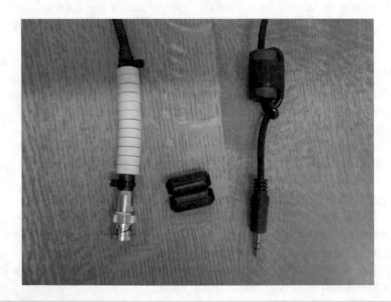

Fig. 8.1 Image showing ferrite collars. (Left) Ferrite collars fitted to coax cable. (Middle) Clip-on ferrite collar. (Right) Power cable; note the cable is passed through the collar and then back through itself

Just cut the cable on either side the ferrite collar, being careful to cut away the insulation on the ferrite collar with a sharp craft knife without scoring the ferrite collar's surface. These collars can then be reused on other cables.

There are other ways to prevent interference within cables. Try not to coil cables tightly in small loops. This is seen time after time, where cables are coiled in small loops and then taped up, the usual excuse being that it looks a lot tidier. This causes problems with power cables, as it can cause them to overheat if heavy loads are being drawn through them. It's a bad idea to do this with coax antenna cables, as this inadvertently makes the cable into an *amplitude modulation (AM) antenna*, worsening the interference since it generates AM signals in the coax cable.

If it is necessary to coil cables, especially coax antenna cables, try and make large loops—the larger the better. The same applies to audio cables. It's a good idea to keep audio cables as short as possible.

Shielding from Computers

Some electrical items are worse than others for producing interference. Electric motors, transformers and high voltage switching equipment are really bad. If problems are encountered with the computer, try reducing the laptop screen brightness or consider turning the screen off if possible. Another strategy is to keep it as far away as possible from the SDR receiver and other sensitive equipment, perhaps 1 m (3 ft) at the very least. A USB extension lead may be needed; choose one with good shielding properties or fit ferrite collars to each end.

Shielding of SDR Receivers

If the SDR receiver itself is fully enclosed in a metal casing, it will be well shielded. This metal enclosure acts like its own personal Faraday cage, protecting it from most interference problems. The only way any stray interference can enter the SDR receiver itself is through the cable connections; this is why SDR's in plastic cases should be avoided.

Just be mindful of where the SDR receiver is placed. Keep it away from strong electrical and magnetic fields, i.e. transformers, and also heat sources, as some internal components of a SDR are affected by heat. A problem with SDR packages is that to keep costs down, a cable with poor shielding is supplied with some SDR's. It's hard to tell just looking at the cables which have good shielding and which do not. Cables with good shielding are usually (but not always) slightly bigger in diameter. This problem is easily solved using a couple of clip-on ferrite collars; just clip one on either end of the supplied USB cable and this usually does the trick. Have the SDR software running as the ferrite collars are clipped on the cable; there should be an instant, noticeable reduction in noise within the signal.

Chapter 9

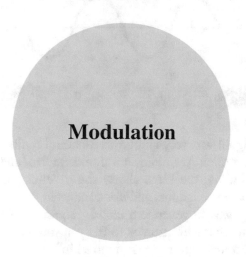

Modulation

Signal Modulation and Signal Identification

While engaging in the projects within this book, we will encounter two types of radio signals. The first is an artificial radio signal, and the second is a natural radio signal. The trick is knowing which is which. We also need to know if these radio signals are modulated or not.

Going back to the first radio transmitters, we can see they are no more than a coil that produces a spark. This is radio in its simplest form. The signals produced from these first transmitters had no particular frequency or wavelength. Anyone with a simple antenna and receiver could pick up the signal. There was no modulation to the signal—it was either on or off. This was enough to send messages in Morse code. This meant that if two ships were sending a message to each other, a third ship might come along and use their radio, causing chaos. The ship with the most powerful transmitter would overpower weaker transmitters and its signal would be picked up.

A quick experiment to try: tune an AM radio so just static is coming out of the speakers. Then holding an electronic gas lighter about 1 m (3 ft) away and pull the trigger on the lighter. A short, sharp "click" sound will come from the radio's speakers. This is how the first radio transmitters worked.

The original version of this chapter was revised. The correction to this chapter is available at https://doi.org/10.1007/978-3-030-54906-0_20

© The Editor(s) (if applicable) and The Author(s), under exclusive license to Springer Nature Switzerland AG 2021, Corrected Publication 2021
S. Arnold, *Radio and Radar Astronomy Projects for Beginners*, The Patrick Moore Practical Astronomy Series, https://doi.org/10.1007/978-3-030-54906-0_9

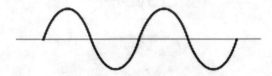

Fig. 9.1 Sinusoidal carrier waveform

The first radio signal was very similar to a natural radio signal—lightning. Do the same thing to an AM radio in a thunderstorm, and every time there is a flash of lightning, the same short, sharp "click" will be heard. For safety, the radio *must* be indoors and *not* connected to any external antenna.

The next thing was to generate a *carrier signal* of a known frequency. This would mean adding electronics to the transmitter to generate the carrier wave. A sinusoidal waveform is pictured in Fig. 9.1.

Almost all carrier waves are sinusoidal, as this shape is quite easy to manufacture and responds well to modulation and demodulation. Now that we have something with wavelength and frequency, a receiver tuned to the frequency of the carrier wave can receive it. A carrier wave on its own cannot carry information, but by varying the length of the time the wave is on or off, the dots and dashes of Morse code can send information.

A carrier wave can be modulated to carry information in three ways. These are: amplitude modulation, frequency modulation and quadrature modulation (IQ), sometimes referred to as digital modulation. Some types of modulation such as amplitude and frequency modulation can be broken down further, either to aid the wave's transmission or to aid its ability to carry information.

Amplitude Modulation

In a modulated radio signal, the carrier wave merely acts as a vehicle to carry other information via the modulation made to it. The first type of modulation is *amplitude modulation* (AM). Frequency is a measure of the number of repeat cycles per second, and amplitude is a measure of the height of each peak. Figure 9.2 shows an amplitude-modulated carrier wave.

The carrier wave can be seen in the middle portion surrounded by an envelope of two sine waves at a much lower frequency. The amplitude intensity of the carrier wave increases and decreases due to the modulation. The modulation in the image is uniform; in reality, it would vary due to

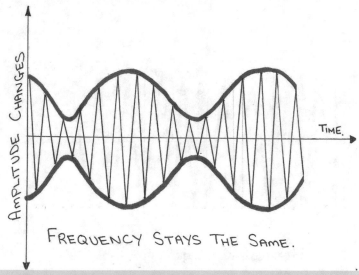

Fig. 9.2 Amplitude-modulated carrier wave

what is being broadcast, either music or speech. An advantage of this type of modulation is that it is easier to demodulate the incoming signal. One disadvantage is that it is prone to interference, which can induce it own modulation on the signal, as mentioned earlier in the case of lightning discharges.

Frequency Modulation

In this type of modulation, the amplitude of the signal stays the same but the frequency of the carrier wave is modulated. See Fig. 9.3.

Anyone who has seen a "slinky spring," the springs that come down steps on their own, has experienced this type of modulation as the coils within the spring opened and closed. The advantage of this modulation is that it is less prone to interference from lightning, etc. However, it has two main disadvantages. The first is that it is harder to demodulate, and more electronics are needed to retrieve the information from the carrier wave. Secondly, frequency modulation doesn't travel great distances. It is fine for local radio stations, but don't expect to pick up FM signals in the Australian outback.

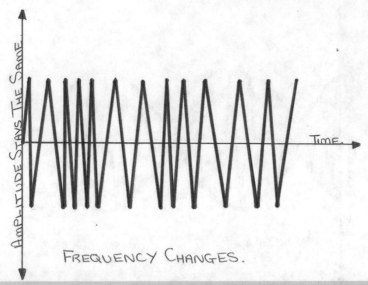

Fig. 9.3 Frequency-modulated carrier wave

Quadrature Modulation

Quadrature modulation requires adding a phase change to the carrier wave. Two signals are said to be in quadrature if they are 90° apart. See Fig. 9.4.

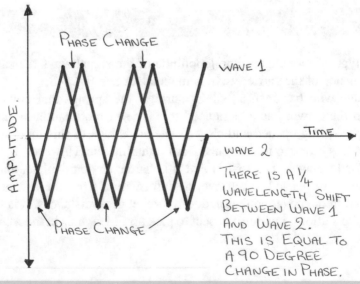

Fig. 9.4 Image showing two signals in quadrature

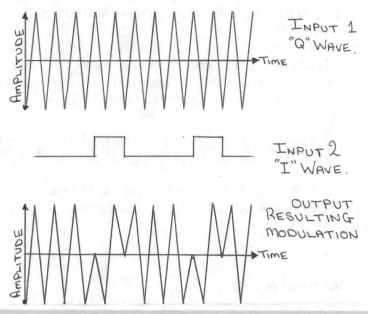

Fig. 9.5 Image showing Quadrature modulation

The first wave is the sine wave (Q). By altering the second input, the cosine wave (I), by a ¼ wavelength, we have introduced a 90° phase change. With amplitude and frequency modulation, we either alter the carrier wave's amplitude or frequency by modulating the carrier wave with the signal we want to send. Quadrature modulation allows us to make more complex modulation changes. We can alter the phasing, amplitude and frequency of the I and Q waves in Fig. 9.4 independently of each other.

To show this in action: if we leave one input as a triangle wave but change the second input to a square waveform to represent a digital input (this need not be digital but this shows the modulation changes well), we can see the resulting modulated waveform. Take a look at Fig. 9.5.

If the phasing input is changed to a digital signal, the carrier wave changes shape to match the input.

Subdivisions of Frequency Modulation

The first subdivision of frequency modulation is *Wideband Frequency Modulation* (WFM). This setting is used when listening to FM radio stations, for example music and news between the frequencies of 88 MHz and 108 MHz. Each station is assigned a frequency, and they are separated from each other by 200 KHz, so there is enough room for 100 stations between 88 MHz and 108 MHz.

The second is *Narrowband Frequency Modulation* (NFM). This setting is used when picking up the International Space Station (ISS), walkie talkie radios and amateur radio operators. The amateur radio frequencies are 144–148 MHz and 420–450 MHz. The bandwidth of NFM is limited to 3.5 KHz.

Subdivisions of Amplitude Modulation

AM amplitude modulation is used for receiving AM radio stations, air traffic control and shortwave radios. It is also used to receive radio signals from Jupiter and the Sun.

Double Sideband (DSB)

Double sideband (DSB) is similar to AM but the carrier wave is suppressed. This has the effect of reducing the bandwidth, but at a cost, as the receiver needs an extra modulation unit to receive the signal.

Upper Sideband (USB) and Lower Sideband (LSB)

Upper sideband (USB) and *lower sideband* (LSB) are used by amateur radio users to transmit voice and data using smaller bandwidths, as only the higher or lower frequency part needs to be transmitted. It is the upper sideband that is of interest here, as it is the one used for detecting echoes from the ISS and meteors.

Artificial or Natural Signals

We can use a few of the tricks that SETI uses to spot the difference between an artificial signal and a natural signal. The first and by far the easiest to do is to check the modulation that the signal is being received on. If it's frequency modulation, then it's definitely artificial, as nothing in nature produces frequency modulation.

The second is to take note of the bandwidth of the signal. Manmade signals are narrowband in appearance, spanning only a few KHz and usually quite powerful. Their frequency doesn't deviate much over time. This is what made the Wow! signal stand out, as it was narrowband and its frequency remained constant for the 72 s it was picked up. It had all the characteristics of an artificial signal.

Natural radio signals are usually broadband and are received at a number of frequencies simultaneously. Take the Sun for example. If the Sun is picked up at 25 MHz, there is a good chance the same signal can be picked up at 20 MHz. A good way to check if the Sun is being received is just to retune the receiver a few MHz on either side. If the same signal is received, there is a good chance it's the Sun.

Another trick is to watch the display. If the signal drifts up to a higher frequency or down to a lower frequency over several seconds or a minute, this is a good indictor that it is natural rather than manmade, as natural signals can drift randomly. Don't confuse this with Doppler shift within the signal, as this is entirely due to natural variations within the received signal.

Unfortunately, manmade interference can sometimes show up as a natural signal. Large industrial electric motors are particularly bad for this, and Wi-Fi routers can be a pain.

Take notes of when the interference is received. Is there a pattern to it? Does it happen at the same time everyday? This is surprisingly effective at eliminating false signals and judging the best times to observe.

The odd pop and crackle may be heard when a receiver is used; this could be lightning from a far-off storm or a neighbor's heating thermostat switching on and off. If used in a radio-quiet area, the dual dipole antenna for picking up Jupiter and the Sun is capable of picking up pulsars. Pulsars, as discussed earlier, can be easily mistaken for artificial signals (LGM's). If one is suspected, take note of the time of day, and using a star atlas or planetarium software, work out what portion of the sky was passing through the antenna beam at that particular time, checking if there are any pulsars in the area.

A good piece of software for this is "Radio Eyes," available from Radio-Sky Publishing at http://radiosky.com. This lists lots of pulsars and their frequencies.

Additionally, look for symmetry in the signals. All artificial signals will show some amount of symmetry to them, whereas natural signals do not have as much. Learning to spot this comes with experience. Finally, the last and hardest thing to check is the polarization. Most manmade signals are vertically polarized, whereas most natural signals are horizontally polarized, but there are exceptions.

As mentioned earlier, during the pursuit of radio astronomy, lots of weird and unexplained signals will be heard. Here is a link to a video clip, https://www.youtube.com/watch?v=TJSctUv5IUc, where a strange signal was picked up. It turned out to be another amateur radio operator who had fallen asleep at their radio!

Part II

**Self-Build Tried
and Tested Projects**

Chapter 10

The Stanford Solar Center's SuperSID Monitor

The SuperSID Monitor

The Stanford Solar Center's SuperSID monitor provides an excellent introduction to the fascinating hobby of radio astronomy, especially for those who do not have much space, as the antenna needed for the SuperSID monitor doesn't need to be placed high up. No antenna mast is needed, and the antenna will work perfectly well at ground level, either indoors or outdoors. In fact, the indoor antenna can work very well.

The SuperSID monitor is self-calibrating, and specialist knowledge is not required to operate the unit. It can be used to produce real science data or simply as an accompaniment to visual astronomy of the Sun. It can also be used in conjunction with the Radio Jove receiver, and by looking at the data, it gives a good indication that a possible aurora is on the way. The SuperSID monitor is used to monitor the ionosphere for *sudden ionospheric disturbances* (SID's) and *sudden enhancement of signals* (SES's). This is done by using the signal from transmitters situated around the world that are primarily intended for a nation's communication with submarines. By monitoring the ionosphere in this way, we can observe the Sun's activity in relation to X-ray flares, coronal mass ejections and other phenomena referred to as space weather.

© The Editor(s) (if applicable) and The Author(s), under exclusive license to Springer Nature Switzerland AG 2021
S. Arnold, *Radio and Radar Astronomy Projects for Beginners*, The Patrick Moore Practical Astronomy Series, https://doi.org/10.1007/978-3-030-54906-0_10

Fig. 10.1 The SuperSID monitor

SuperSID monitors are inexpensive to buy and require minimal DIY skills to build an antenna. The size of the SuperSID monitor is no larger than a shoe polish tin, measuring 80 mm (3.25 in.) in diameter. It requires no soldering or electronic work at all, except for the fitting of a screw-on BNC connector and the coupling of wires into a jointing box using a screwdriver. See Fig. 10.1.

The SuperSID monitor comes with its own power supply in the form of a transformer. If the monitor is being used in the United Kingdom, an adaptor will be needed; a two-pin electric shaver plug is just fine for this purpose. It also comes with the required fittings, although it will need a short length of RG 58 50 Ω antenna cable, available from any electronic shop. Wire will be needed to make the antenna itself. A minimum of 120 m (393 ft) of 22 AWG (American Wire Gauge) or 23 SWG (Standard Wire Gauge) will by fine; this too can be bought from an electronic shop.

The wire will be wrapped around the antenna frame after its construction to form a working antenna for the SuperSID monitor. The construction of a suitable antenna will be covered later in this chapter. Ordering the kit over the internet is easy, from either the Stanford Solar Center or from the Society of Amateur Radio Astronomers (SARA) website. It should arrive within a couple of weeks. If ordering for delivery into the United Kingdom, there is the usual import duty to pay and extra postal charges will apply.

On unpacking the parcel, you will find included with the SuperSID monitor a CD-ROM with the operating program. As of the time of writing in 2020, a CD reader on your laptop or desktop is still required. See the list of requirements here: http://solar-center.stanford.edu/SID/installation/. Other information, such as a printable manual, is also included. It is handy to have a hard copy of the manual to browse through. There is also a printed sheet with the observer's own personal monitor ID number, plus the site name that was chosen before ordering. Keep this safe as it will be needed later. This information is used if you wish to take part in research, as the programme can be set to automatically upload the SuperSID data back to the Stanford Solar Center, where it is collected and processed from observers around the world. This is optional, and everything will work just as well if you choose to be a standalone observer.

There is an email address in case any problems arise, where a member from the Stanford Solar Center will be able to answer any questions or offer advice. They are very quick when it comes to replying and in most cases an answer is received the following day.

There are a number of other space goodies included with the SuperSID monitor. These are a NASA Sun fridge magnet, a 3D postcard of the Sun, a 3D NASA bookmark, a large 3D poster of the Sun, and a DVD entitled *Cosmic Collisions*, along with leaflets and a sticker for SARA (Society of Amateur Radio Astronomers). See Fig. 10.2.

Once the SuperSID monitor is up and running, no further input is needed, apart from checking the condition of the antenna every now and then if it is kept outside. The monitor is quite happy just doing its job recording data; all the user need do is check the data for evidence of solar activity.

Using the SuperSID monitor to observe the Sun every day, you will notice changes in the ionosphere directly overhead. These changes can be seen from month to month and season to season. Once you are familiar with the data, this project is very easy and rewarding.

Space Weather

The Sun has been throwing out a constant stream of charged particles, known as the *solar wind*, ever since the first day nuclear fusion started within its core millions of years ago. The intensity of the solar wind has a lot to do with the 11-year solar cycle, where the polarity of the Sun's magnetic field flips.

At solar minimum, the magnetic field lines enter the magnetic poles of the Sun. These magnetic poles are close to the axis of rotation of the Sun.

Fig. 10.2 SuperSID parcel unpacked with other space goodies

Midway between the flip is what astronomers call the solar maximum. The magnetic field lines are now at right angles to the Sun's axis of rotation. With the Sun being a gaseous body and having no solid surface, it rotates at different speeds at the equator and the poles. This twists and distorts the magnetic field lines. For example, imagine a garden hose that instead of being neatly coiled on a reel is just thrown in a heap. The hose will soon become kinked and knotted. The tension builds up in the field lines like giant elastic bands and at some stage, something has to give. The magnetic field lines snap and can break through the surface, producing sunspots and other solar outbursts. Sunspots sometimes come in pairs, with one of a north polarity and the other of a south polarity, just like a magnet.

When magnetic field lines break, a large amount of energy is released, and this is known as a *coronal mass ejection* (CME). This energy travels through space in two particular forms, one as a high energy electromagnetic wave traveling at the speed of light that reaches the Earth in a little over 8 min, and the other as high energy charged particles within the solar wind that can travel at speeds of several thousand kilometers per hour. These can take, depending on their velocity, somewhere in the region of 2 or 3 days to reach the Earth.

This type of solar activity has been happening for millions of years, so why bother about it now? The reason is that we have crewed spaceflights. If one of these huge CME's explodes on the Sun, the astronauts will be in trouble, as all they will have to protect themselves is the spacecraft itself. In a little over 8 min they will be exposed to huge amounts of ultraviolet radiation and X-rays equivalent to thousands of medical X-rays. If they somehow managed to survive this, in a day or so they would be hit by the high energy particles within the solar wind, which would have so much energy that they could pass straight through the spacecraft, including the astronauts' inside. The risks would be even greater with a crewed mission to Mars, as the round trip would be in the region of 3 years.

It isn't practical to carry the lead shielding needed to protect the astronauts against these types of radiation, and the best idea that scientists have come up with at the moment is to use water—several tons of it—as a protective screen. This would be achieved by having a room in the centre of the spacecraft with water surrounding it. The crew would have to stay within the confines of this one room until the threat overtakes the spacecraft. Depending on the distance at which the crew is from the Earth, the type of threat involved and the speed at which it was traveling, at best, we could give the crew maybe a couple of days of warning in which to prepare and take the necessary precautions.

Here on Earth, we can be at the mercy of high energy radiation through our reliance on satellites. Satellites, in orbit above the protective atmosphere, can be subjected to high energy radiation from the Sun. These include the global positioning satellites and the cellphone network of satellites. The high energy radiation can overload their delicate electronics and literally "fry" them, just like placing a cellphone in a microwave oven.

This could lead to all manner of navigation and communication problems, and could prove very expensive for businesses such as international banking, which rely on satellites.

There is another problem that can affect thousands of people. All over the world we have high voltage power lines that carry electricity from the power stations to our homes. They can stretch for hundreds of kilometers at a time. These power lines can act like a huge antenna and collect the high energy radiation, which can induce an electrical current to flow within the wires of the power lines, causing them to overheat, or even worse, to overload. This can be a real problem for the transformers that bring the high voltage down again for use in our homes, as these too can overload and burn out.

The only option open to the power company is to shut down the area that is likely to be affected by such radiation, but this is not as simple as just turning off a light switch. All this is not too much of a problem for hydroelectric generation, where valves can be shut to cut off the flow of water. But for turbines using steam to generate the electricity, it can take days to slow down before the turbines can be stopped, so the power needs to be diverted away from the affected area. There is the added problem of vital services like hospitals and the other emergency services, not to mention the domestic consumers who would not be too happy going without electricity.

All these services rely on accurate space weather forecasting. A warning is normally issued when X-ray solar activity with a magnitude of M5 is detected. (X-ray classification will be covered later in the chapter.) It can be seen that there is a genuine threat on human activities from the high energy radiation from the Sun.

How the SuperSID Monitor Works

Situated around the globe are powerful transmitters that different countries use to communicate with their submarines. The antennas for these transmitters are huge, measured in kilometers, due to the wavelength at which they operate. As we have already discussed, longer wavelengths are better at penetrating the deepest oceans of the world. These transmitters are so powerful that their signals can travel around the Earth, bouncing off the ionosphere. These frequencies have the advantage of being very stable.

The SuperSID monitor doesn't receive the communications like a radio receives music. No communication between the submarines and the bases on land can be heard. Any communication would be in code anyway, for obvious security reasons. We are only interested in the carrier wave frequency from these transmitters.

As discussed in Chap. 9, a carrier wave in its simplest form is a signal frequency wave that is used to carry information, and can be modulated by amplitude, frequency or digitally coding. The signal from these powerful transmitters bounces between the Earth and the D layer of the ionosphere, and by doing this it can travel around the world. The SuperSID monitor picks up these signals through its antenna and amplifies them to give a usable signal for recording and processing on a computer. If there is any change in the ionosphere, for instance as the result of a solar X-ray flare from the Sun, this can cause a sudden drop in the signal strength. These are the sudden ionospheric disturbances we mentioned earlier.

Solar flare activity can also cause a sudden increase of signal strength, and these are referred to as SES's—sudden enhancement of signals. The SuperSID monitor will pick up this change in signal strength. This is only possible between the observer's local sunrise and sunset times, as the ionization in the D layer is lost at night, and the monitor can no longer function without it. This means only the data from local sunrise to sunset needs to be checked.

If there is a large solar flare, the monitor can sometimes be overwhelmed and the signal may be lost completely. This doesn't harm the monitor, but it will show up on the graph when processed later. Sometimes a transmitter signal will be lost for a short time and the graph will just show a flat line. If there has been no solar activity to overwhelm the monitor, this could indicate that the transmitter has probably been shut down for a short time in order for maintenance work to be carried out. It has been suggested that by taking note of the time and date of the transmitter shutdown, a pattern of the transmitter's maintenance can sometimes be worked out.

A list of suitable transmitters is supplied with the SuperSID monitor, or from the Stanford Solar Center website. Each observer will choose their own transmitter(s) on the basis of signal strength; this process will be covered later. The SuperSID monitor is designed to monitor six different transmitters simultaneously, and it takes a reading every 5 s throughout the day. At the end of a day, using the software provided, you can produce a graph, which can then be examined for traces of solar activity.

Look at Fig. 10.3. This is a portion of a graph starting at 8 am and ending at 6 pm, showing two changes within the signal that the SuperSID monitor has recorded.

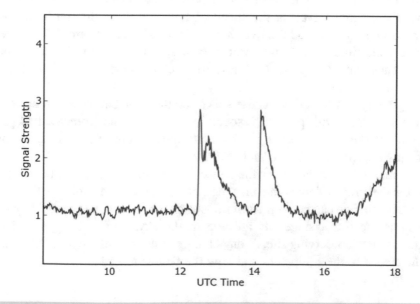

Fig. 10.3 SuperSID data showing two flares

These are X-ray flares and will be covered more in-depth later, but for now, look at the peaks. The line on the left of each peak is vertical. This shows that the ionization of the ionosphere happened almost instantly, whereas the right-hand side shows that the ionization decayed to background levels over nearly 2 h for the first peak and nearly an hour for the second. This image shows the classic shape to watch out for, although the peaks could go down or up depending on whether it is a SID or SES change in the signal. The height of the peak can also change depending on the intensity of the flare. Additionally, some flares can last an hour or so, like the ones above, while others can last all day.

For those who are interested in using the SuperSID monitor with other forms of solar activity, it's a good idea to sign up to an aurora warning service. There are plenty available, and a search online will show a number of them. It is important to sign up to a local one, or at least one within the local time zone, as this makes the information sent more relevant. These aurora warning services automatically send out an email or text to let recipients know that there is a high probability an aurora may be observed over the next few nights.

If a large peak is picked up within the SuperSID data, the recipient should keep an eye on their emails because it is likely to trigger an aurora warning. This is due to the fact that at the time an X-ray flare is released from the Sun, there also will be a vast increase in the release of the high energy particles that cause the Earth's auroral display. The X-ray flares burst from the Sun with tremendous power. The electromagnetic energy from these flares travels at the speed of light, reaching the Earth and our SuperSID monitors in a little over 8 min. The high energy particles released at the same time travel at a slower rate of several hundred kilometers per hour. They can take up to 2 or 3 days to reach the Earth, so the aurora display will be delayed.

Lightning can sometimes cause spikes on the data, but these can be spotted quite easily with practice, especially if the weather forecast has predicted thunderstorms, although thunderstorms don't have to be local events for the SuperSID monitor to pick them up.

As mentioned earlier, over the space of a year, changes will be noticed within the graphs depending on whether or not there is solar activity. A dramatic change can be seen from midsummer to midwinter, caused by the fact that the Sun changes its altitude in the sky. These changes will be unique to each observing site, as any change in latitude of the Sun will make a difference to the equipment and how the data is recorded.

How to Make an Antenna

The antenna for a SuperSID monitor is a simple wire-wound antenna and can be made from a variety of materials such as wood, plastic or a mixture of the two. The shape of the antenna isn't too important; the simplest design is a wooden cross. See Fig. 10.4.

Figure 10.4 shows an antenna that has four spokes. Each spoke is 705 mm (27.8 in.) in length when measured from the centre. This gives a total area of 1 m squared. The 120 m (393.7 ft) of 22 AWG 23 SWG wire is then wrapped around the edges to form a square measuring 1 m (39.37 in.) on each side. Solid core wire is preferred over stranded core wire, as the solid core wire holds its shape better since it is less flexible than stranded wire. Stranded wire can be used, but it would need more support.

This is the smallest recommended antenna size and possibly the easiest to make. It doesn't have to be square—it can have six, seven and even eight sides, or even be circular. It all boils down to the room available to place the antenna, and your practical DIY skills.

The size of the antenna has a lot to do with location. For those living in a location with a high-strength signal, this small antenna would work fine. But on the other hand, a larger antenna would give more gain and would possibly provide more stations to monitor.

Fig. 10.4 A simple SuperSID antenna

The pictured antenna was the first antenna I built; it had 200 m (656 feet) of wire wrapped in 50 turns around the frame, giving a total area of 50 m². It is better to have 25 turns on a 2-m (72-in.) frame than 50 turns on a 1-m (39.37-in.) frame. A 2-m (72 in.) frame with 25 turns would give an area of 100 m², four times the size of a 1-m antenna.

This first antenna produced a signal, but not one of a suitable strength. Within the SuperSID manual, it is recommended that adding more turns of wire will improve the sensitivity, so another 100 m (328 ft) was added. This was an extra 25 turns to the frame, which gave a total of 75 m², and although this helped, the signal strength still wasn't high enough. So, an additional 100 m (328 ft) were added, making 400 m in total, with 100 turns of the antenna and 100 m² in area. This created a new problem, as the sensitivity had increased but the resistance within the wire had also increased, so any benefit from the added wire was cancelled out by the resistance within the wire itself. The only option was to build a larger antenna frame. See Fig. 10.5.

This particular shape was used because it gave more area than the smaller square antenna and also followed the shape of the inside of the roof. Each

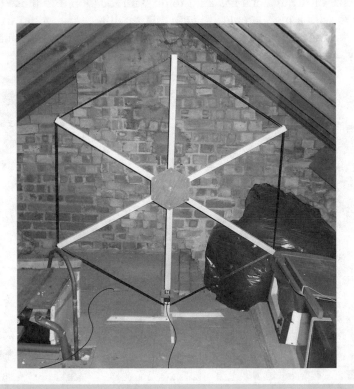

Fig. 10.5 Improved SuperSID antenna, sited inside attic space

spoke is 750 mm (29.5 in.) when measured from the centre, and this gave a total area of 75 m² using only 200 m (656 ft) of wire. This led to a marked improvement of the signal being received and was enough for the SuperSID monitor to function correctly. As can be seen from the image, the antenna was the largest that could be accommodated in the location where it was to be kept. But larger antennas can be used if there is room.

To make a simple 2-m square antenna, the following will be needed:

1. Two pieces of wood, with a 50-mm (2-in.) square section, 2.8 m (111 in.) in length
2. An 8-mm (0.31-in.) bolt, some 125 mm (5 in.) in length
3. Assorted screws
4. Electrical tape
5. Wood glue
6. A small amount of 6-mm (0.25-in.) plywood to support the centre
7. Enough wood to make a simple stand to support the finished antenna
8. Wood preservative and/or paint if the antenna is to be kept outside

Start by cutting a joint in the centre of the two lengths that will make the frame of the antenna. See Fig. 10.6.

This will allow both pieces to fit at 90° to each other. Try and keep this joint as tight as possible because this is going to take the weight when winding the antenna.

Next, cut a slot in the end of each of the four ends to accept the wire. See Fig. 10.7.

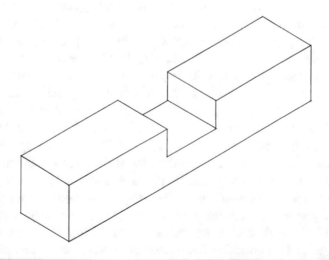

Fig. 10.6 Central joint of frame (not to scale)

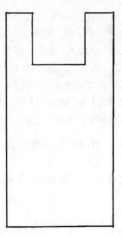

Fig. 10.7 Slot cut ready for wire (not to scale)

When these have been cut, a good tip to avoid problems later is to round off the inside of the slots so that the wire fits around a curve rather than a sharp edge. This will prevent the wire from suffering any undue stress, especially if the antenna is to be kept outdoors where movement by the wind could cause the wire to break. See Fig. 10.8.

Now it is time to assemble the antenna. First glue and screw the two lengths together at the central joint. The screw is only there to hold the joint together while the glue dries; it is best to leave this to dry overnight. Then, cut two pieces of 6-mm (0.25-in.) plywood into pieces 300 mm (12 in.) square, to be used to support the centre of the frame. Remove the centre screw when the glue has dried and place a square of ply on each side of the frame over the centre for support (See Fig. 10.5), glue and screw these on.

Cut a small piece of 6-mm (0.25-in.) plywood approximately 50 × 75 mm (2 × 3 in.), and screw this to one of the spokes about 75 mm (3 in.) from the end. This is for the antenna cable jointing box. Build a stand to support the antenna. This can be any design; see Fig. 10.9. This is a simple design, and no joints need to be cut. This is the type used in the previous figures and can easily be made using the same-sized material used for the antenna frame. The height of the stand need only be just over half the size of the finished antenna, and the 4 feet would be quite adequate to hold the weight of the antenna. If you are leaving the antenna outside permanently, a more solid support will be needed, or spikes could be driven into the ground through the 4 feet, after first checking for power and telephone cables, plus gas or water pipes.

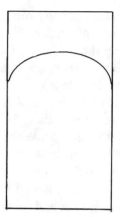

Fig. 10.8 Inside of slot rounded off to protect wire from sharp corners (not to scale)

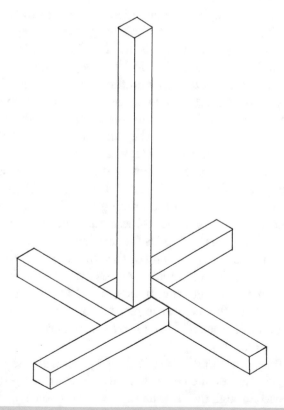

Fig. 10.9 A simple design for an antenna stand (not to scale)

When everything is dry, clean off any rough edges with sandpaper and apply the paint, varnish or wood protection of choice. It is now time to wind the antenna. Start by drilling through the center of the antenna frame to accept the 8-mm (0.31-in.) bolt. Lift the antenna frame against the stand and leave enough clearance so that the antenna can be turned without catching the bottom of the stand. Drill a corresponding hole in the frame support. Now, slide the bolt through the frame, then slide two washers onto the bolt and lift the antenna frame into place. Slide another washer on the bolt and loosely tighten the nut. It should now look like a rather crude windmill with its blades. Fix the jointing box to the frame to ready for fixing the wire and antenna cable. The wire in the United Kingdom comes on 100-m (109-yd) reels; two reels are needed. Join one end of the wire from one reel to the other, preferably by soldering and then covering the joint with electrical tape. Now wind both reels onto one. There should be room on one reel for all the wire. Cut several lengths of electrical tape into 50-mm (2-in.) lengths and stick these somewhere handy, like the edge of a table, as this will save you some struggling later.

Eyelets for the jointing box are supplied. Fix one to the end of the wire either by crimping, or better still, by soldering. We are now ready to wind the antenna. First, couple the eyelet to the jointing box, leaving a little bit of slack wire, and apply a little electrical tape to the antenna frame to hold it in place. Ask an assistant to help with the winding by holding the reel of wire and applying a light pressure to act as a brake and stop the wire from becoming knotted or tangled. Avoid kinking the wire, as this will weaken it.

Start to turn the antenna and lay the wire in the slots at the end of the spokes. Keep a steady tension on the wire, as this will help it support itself. Apply a little electrical tape every now and then to help keep the wire in place, and carry on until the end of the reel is reached. Take it to the nearest full turn back to the spoke with the jointing box. Leaving enough wire to make the connection, wrap electrical tape around the spoke to hold the wire from coming undone. This concludes the antenna winding.

Now fix the other eyelet to the end of the wire and fix it to the other connection on the jointing box. If a multimeter is available, set it for resistance and touch the two ends of the wire. If the reading is about 10 Ω or less then the antenna is fine, as this is just the resistance within the wire. If there is no reading, then this means there is a fault somewhere, like a poor connection or broken wire.

Assuming that everything is fine, put the spoke with the jointing box at the bottom. Then, tighten the center bolt to hold everything in place. It is a good idea to drill through the antenna support into one of the spokes and fit a screw to hold the antenna solid and stop it rotating and damaging the

cable. A spacer will be needed to take up the space of the plywood support-
ing the centre of the frame; this can be done with washers or an off cut of
the same plywood used earlier.

Next, fix the coaxial antenna cable to the jointing box. Before stripping
the insulation off the antenna cable, it is a good idea to fit two or three ferrite
collars to the end of the cable that is coupled to the antenna itself. These just
slide onto the cable, and after the connection is made they are slid back up
as close to the connection as possible and held in place with electrical tape
or a cable tie. These collars will help stop interference involving radio frequencies entering through the open end of the antenna cable. See Chap. 8
for more details.

Next, start by stripping off the outer insulation of the cable, being careful
not to damage the braiding below. One or two broken strands of braiding are
inevitable and fine, so long as it doesn't run to half a dozen or more.
Carefully, using a small screw driver or something similar, untangle the
braiding. It's a good idea to use the tip of a ballpoint pen for this, as there
are no sharp edges. When this is done, twist it together to form a single wire.
Next, strip the insulation off the centre conductor, but no more than needed
to make a connection. As before, be careful not to damage or score the central wires. Crimp or solder on the other two eyelets and couple them onto
the jointing box. It doesn't matter which way around they are.

Leaving a little slack cable, apply a plastic cable tie or electrical tape to
the cable to take the weight off the connections. If the antenna is going to
be used outside, be sure to apply enough electrical tape or other water-
proofer to protect the connections. Waterproofing of antenna cable is covered in the Radio Jove Chap. 12.

Next, fit the BNC connector. To do this, mark off 20 mm (0.75 in.) of the
outer insulation, being careful as before not to damage the braiding below.
Fold the braiding back over itself. Then strip about 6 mm (0.25 in.) off the
central insulator and make sure the central conductor has no stray wires.
Twist any strays together.

Now, turning the BNC connector clockwise, twist it over the end of the
antenna cable. It should only need hand pressure to fit in, so don't be
tempted to use pliers to force it on. After the BNC connector is fitted, a good
idea is to wrap one or two turns of electrical tape around the spot where the
BNC connector meets the antenna cable's outer insulation, as an added
precaution to keep out damp and dirt.

Try and keep the cable from the antenna to the SuperSID monitor as short
as possible, as this makes a difference to the signal. The recommended
length is 10 m (32.8 ft), but 8 m (26.24 ft) will do. The antenna is now ready
for use. For more about different antenna designs, visit the Stanford Solar
Center website to explore all manner of innovative and clever ideas.

Sighting the SuperSID Antenna

Now that the antenna has been made, it must be sited in a place where it can perform its duties. This doesn't have to be high up, as the antenna will work at ground level. First, the obligatory safety warnings:

1. The antenna is likely to be bulky and heavy, depending on the design, and assistance may be required to maneuver it.
2. Don't mount the antenna anywhere it is likely to cause an obstruction.
3. Check that the antenna cable cannot be tripped over.
4. If mounting the antenna indoors, make sure to take account of the locations of any power cables or water and gas pipes before drilling into a wall.
5. If working at height, for instance off a ladder, be sure to ask for assistance from a reliable person.
6. If the antenna is to be mounted outside, make sure any fittings are secure and there is no danger of the antenna falling, especially if it is windy.
7. Don't fix the antenna higher than surrounding buildings or trees because of the risk of a lightning strike.
8. In fact, in the event of a local thunderstorm, unplug the antenna from the SuperSID monitor until the storm passes.

The antenna must be placed as far away from electrical interference as possible. This can be easier said than done, as electrical interference is likely to be everywhere. It is more of a case of finding a location with a tolerable level rather than finding an interference-free location. Some questions to address if the antenna is being used indoors are: Is the building a wooden or brick structure? Does the building contain a large amount of metal or a metal-framed structure? If using the antenna inside a wooden or brick structure, it should work fine, but if inside a metal building or metal-framed building, there is likely to be a problem of interference from the metal structure.

Some people have reported problems with interference from outside lighting, in particular from street lighting. This hasn't been a problem here in the United Kingdom, as the low-pressure sodium lights have been changed to the more efficient high-pressure sodium lights, and white LED lights are replacing these. These cast the light downward rather than skyward. It could be that interference from other sources have overshadowed or masked this problem. The old computer monitors were a problem, but the newer TFT monitors seem to create less interference.

Fluorescent lights have also been reported as a problem, especially the older ones, but more modern fittings have reduced this. Also, touch-sensitive lights, the type that come on at different levels of brightness and are normally used for bedside lights, can cause a certain level of interference. It is much better to have the simple on/off switch type.

Microwave ovens are a real problem, so be sure to keep the antenna as far away from them as possible. One further item that has proved to be a real problem is the computer printer. When printing, it drives the SuperSID monitor's display wild. This is easily sorted out by only using the printer after sunset.

If there's an interference problem and its origin is not known, switch off every electrical appliance in the home, and then switch on one at a time and take note of the display. This may sound a little over the top and time-consuming, but it works. All that said, it is still possible to run a successful setup indoors as long as potential problems are kept in mind.

My first plan for the SuperSID monitor was to keep the antenna in the spare bedroom with the laptop, but this idea was dashed thanks to the microwave oven in the downstairs kitchen (which is directly below the spare bedroom). By simply transferring the antenna to the attic and drilling a small hole through the ceiling for the antenna cable, this put more distance between the offending microwave oven, which allowed the SuperSID monitor to work correctly. This had the added advantage that by being in the attic, the antenna was less likely to be knocked or moved. The antenna was also briefly tried outside, but in this case, it picked up interference from the surrounding houses. Everyone's situation is different, but by persevering and experimenting, a location where the antenna will deliver a working signal can usually found.

Sound Cards

The SuperSID monitor comes with its own operating program. The antenna detects the very small signals that are then amplified by the SuperSID monitor to produce a signal of a suitable strength to work with. These are then fed into the sound card on the computer. The job of the sound card is to detect the information within the inputted signal and process this into information that the operating program can use and store digitally for inspection and processing later.

Any change in signal strength will be detected and registered as a possible SID or SES event. We have already talked about the type of signals that we will be using—the communication transmitters that different nations use to communicate with their submarines. These are very powerful transmitters with many megawatts of transmitting power, but depending where in the world the SuperSID monitor is, some stations will be of more use than others. The sound card allows a greater number of stations and frequencies to be monitored simultaneously, meaning there is a better chance of picking up changes within the ionosphere through any changes detected in the signals. Those living in Europe or Asia can use a sound card rated at 48 KHz, but those living in America will need a sound card rated at 96 KHz in order to pick up the stations.

To find out the details of a sound card on a Windows operating system, click on "My Computer," then click the "Open Control Panel." Once there, look for "Audio Manager" and click this tab, wherein the details of the sound card should be displayed. For desktop computers with a tower unit, it is possible to change the sound card for one with a higher sampling rate. This however is becoming less of a problem now, as computer technology is increasing and processor speeds are getting faster, meaning the quality of the sound cards has increased along with everything else.

Software

Before loading the software, have the paperwork that comes with the SuperSID monitor at hand, as this information contains the monitor number and the site name chosen before ordering. The observer needs to know their latitude and longitude and the time zone in relation to Universal Time (UT). The sampling rate at which the sound card will be operating at on the chosen computer needs to be known.

Insert the disc with the operating program into the computer and it will probably start automatically. If it does not, right click on the disc drive short cut and click "Open." The software needs the above information before the program is run for the first time. Click on the "Configure File" icon. This will open the "Parameters" window, where the specific information of the chosen site needs to be entered. The window displays in a basic notepad format, and the only changes required are: the site name, longitude and latitude, time zone information, monitor ID number and sound card sampling rate. Leave everything else at the default settings.

All this may sound difficult, but it is no harder than programming a destination into the satellite navigation device in the car.

It's a good idea to print off the pages of the instruction manual that relate to the installation of the operating program first, as this provides a visual guide and handy reference. At this stage, station frequencies will not be known and can be left at default settings. The automatic upload setting, which sends SuperSID data back to Stanford, is set for "No" by default. If you wish to send off information later on, it's a simple matter of just clicking this box to change it.

To start the program, it is necessary to access the SuperSID file. Navigate to and click on "superSID.exe," or beforehand, place a shortcut to this file on the desktop. This can also be done with the configure file so this too can be accessed more easily to change station information and set frequencies when the best station for the location has been found. This will be covered in the next section.

If all goes well and the instillation is a success, double click on the shortcut icon to start the operating program. It should look something like the screenshot in Fig. 10.10. Note the shape of the graph; although this is far from an ideal noise-free one, it should give you a general idea of what to look for.

Another point worth mentioning about the program is that it is quite happy doing its own thing in the background and does not interfere with the other programs on the computer. Just drop it down on the toolbar and use the computer as normal, except when using a printer or scanning equipment.

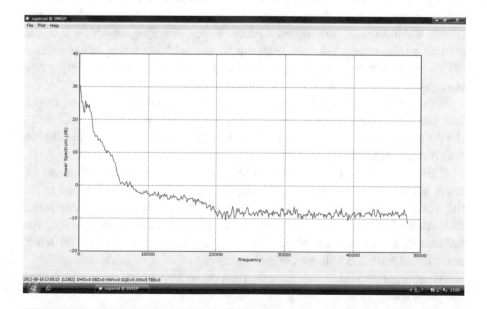

Fig. 10.10 Screenshot of SuperSID's operating program after the software has been installed

Connecting Everything Together

The time has come to connect all the parts together.

Connect the antenna via the BNC connector to the SuperSID monitor. This is a simple bayonet fitting, so a push-and-twist clockwise movement will do the trick, remembering to keep this cable as short as possible, but not to have it at full stretch, or the cable will suffer from undue stress. It was found that removing a short length of cable, such as 1 m (39 in.) can make a big difference to the signal.

Next, plug the SuperSID monitor into a power supply. An adaptor may be needed for this, depending on the design of the domestic electric power sockets. The next part is to plug the jack plug into the computer. Don't be tempted to use any type of extension cable such as the sort used to extend the cable of a pair of headphones, as it won't work and the power of the signal will be lost.

Coupling things to a computer can be a little bit of a minefield, as each computer will have its own fittings and operating system. Plug the jack plug into the inline input socket if it has a separate one, but note that this can sometimes switch on the built-in microphone. If this is the case it will be necessary to access the audio properties on the computer and examine the settings for the microphone input to ensure that the microphone is not live and recording all background sound. Some computers share the same input socket for inline and microphone input. When plugging anything into the socket, an on-screen window pops up to ask which—input either micro-phone or inline input—is required. The computer may ask what sample rate is required from the sound card's operation.

If there are any problems with the SuperSID's operating program, check that the sound card sample rate matches the sample rate that is set within the parameter window, as discussed earlier. This will in most cases probably sort out any problem.

If there is no inline input on the computer, the microphone input can be used, but this is a far more sensitive input and caution will be needed to set the volume input level. If the input volume is set too high, it will cause distortion of the input, in a similar way to a speaker distorting music if the volume is turned up too high. If the microphone input has to be used, experiment with the volume control in order to find a suitable setting.

Assuming everything has gone as planned so far, the next thing is to start the program.

Double click on the shortcut and wait a second or two for a screen to appear. In the bottom left-hand corner, "Waiting for timer..." will be seen, and a second or two later the graph will appear.

If you are lucky, some large, narrow peaks will be seen. If not, don't worry. These peaks can be one of two things: interference which we don't want, for example from a television set, or the stations that we do want. At this stage, it will not be possible to tell which is which. If you are lucky enough to have a peak, then place the computer curser at the top of the peak and left click. In the lower left of the screen, three important pieces of information will be shown: frequency, power and strength. Doing the same thing on a different peak, the frequency, power and strength will have changed. This is how to find out what frequencies give the signals and therefore what stations can be received. These can then be set in the parameter window.

If there are no peaks, or even if there are, it is a good idea to position the antenna to find the best reception. This is done by turning the antenna about its vertical axis and at the same time looking for the tallest peaks above the background level of noise. This is where the help of an assistant can be useful. Turn the antenna about its vertical axis by only a few degrees at a time. It is very important to wait at least 10–15 s for the operating program to register the change. Remember, it only takes a reading every 5 s, so it needs the time to register the movement of the antenna. Carry on turning the antenna like this, and if any large peaks are seen, click on the top of the peak and jot down the frequency, power and strength. It is also a good idea to mark the floor in some way to denote the direction of the antenna at that particular moment. Something as simple as a chalk mark will do. After completing one whole revolution, it should be possible to have an idea where to position the antenna to get the maximum gain from the signals.

If no signals are detected, backtrack and work through each step again, double checking that nothing has been overlooked. If there is still no signal, either find a quieter site with less interference to use the antenna, build a larger antenna, or both.

Assuming there is a signal, place the antenna to get the maximum gain from the signal. The thinner and taller the peak, the better. The old saying "quality over quantity" is true here. Although the SuperSID monitor is capable of monitoring six individual signals simultaneously, it is better to have two strong signals rather than four mediocre ones. Even if only one good signal is detected, the SuperSID monitor will still work.

Make a more permanent mark on the ground to save going through the turning process again, or worse, to protect from some over-curious person moving the antenna in order to admire the craftsmanship of its construction.

The next part of the process will take 2 or 3 days to perform. In earlier chapters we discussed the ionosphere and how during the day it becomes ionized. This is how the SuperSID monitor works, monitoring the changes to the signals reflected back from the ionosphere. The ionization is at its greatest during the hours either side of local midday. It is therefore a good idea to take note of the peaks that are visible around midday. Keep a notepad at the side of the computer, and each day, say, over the weekend, look at the screen at midday and observe which peaks are the tallest. Click on these peaks and jot down the frequency, power and signal strength. If it is possible to do this for more than two consecutive days, maybe over the course of a week, you will see a pattern emerging that shows the strongest signals.

See the screenshot in Fig. 10.11 to get an idea of what to expect. Although again this is far from an ideal graph, it still works well and flares as small as C1.5 can be detected on a very good day.

Within the SuperSID manual or via the Stanford website, a list can be obtained of VLF stations and their operating frequencies and location (latitude and longitude). Each station is denoted by a three-letter ID. For example, GBZ is the ID for the Authorn station used by the United Kingdom, and its operating frequency is 19.6 KHz.

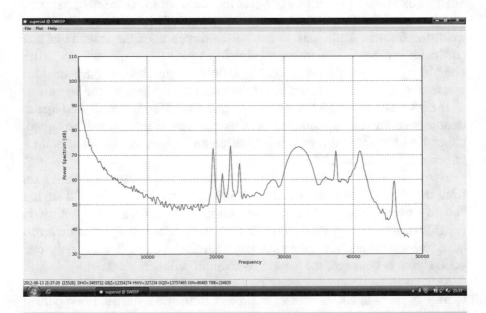

Fig. 10.11 Screenshot to demonstrate usable stations. The four peaks are just left of center

Looking at the data that has been collected over the last few days while running the SuperSID, you can check the frequencies against the list of VLF station frequencies. These are the stations that the equipment can receive, and it may be surprising to see the distance that the signal has traveled from the transmitter to the equipment.

Next, tell the operating program of the SuperSID which frequencies to monitor. This is done by accessing the parameters information again. Once the parameter file has been opened, scroll down until Station 1 is seen on the list. This can be changed in the same way that the site name and location were changed earlier.

Below the heading of Station 1 is a list of three items: (1) call sign, (2) color and (3) frequency. At the side of the call sign, enter the three-letter station ID, for example GBZ. The color can be left alone. "r" for example stands for the color red, and this means that this station will show up on the finished graph as a red line. This is useful when comparing more than one station at a time. If averse to the color red for whatever reason, maybe due to color blindness, you can change the color setting.

Beside the frequency, enter the station's frequency. This must be typed in full, for example as "196000" and not "19.6." Do this for each of the stations on the list, then close the parameter file.

To confirm that the stations that have been programmed into the SuperSID's operating program are in fact stations and not just an annoying bit of interference, there is one further test to perform. Run the SuperSID monitor for at least 24 h to get a day's worth of data. When this has been done, this data should be accessed while the SuperSID program is running. To access the data, click on the plot tab in the top left-hand corner. The list of data for each of the stations that have been selected will be there, and the data can be recognized by the three letters of the station's ID. Click on the required station and open it. A graph will be produced. Look for sunrise and sunset changes within the graph. See Fig. 10.12.

If a graph similar in shape to Fig. 10.12 is produced, this means a station has been found. Remember that it doesn't matter what the graph looks like at night when the ionization in the ionosphere is lost. We are only interested in what happens between sunrise and sunset, as this is where any activity from the Sun will be found.

Other stations can be viewed at the same time by clicking on the plot tab again and clicking on another station. Each graph will be shown in a different color. It is a good idea to view the stations individually until accustomed to recognizing what to look for.

If you are unlucky enough not to have found a station, but some source of constant interference instead, do not worry. Try to eliminate that particu-

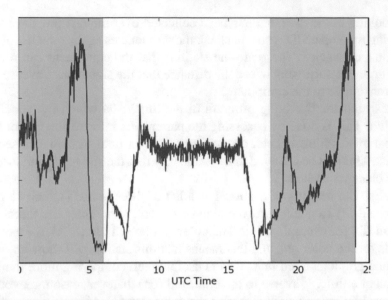

Fig. 10.12 SuperSID data showing sunrise and sunset changes

lar source of interference by turning the antenna until the interference is gone, or has at least been reduced, as long as it doesn't affect other stations. If this is not possible, then it is maybe a case of living with the interference. The frequency of the interference should then be removed from the list of stations within the parameter file and the program will no longer save this data.

Again, once happy that the equipment is functioning correctly, it is a good idea to permanently position the antenna in the preferred orientation, or at least make permanent marks somewhere on the ground so that if any maintenance work is carried out on the antenna, it can be placed back in the correct orientation.

One last note: the SuperSID monitors input signal can be quite interesting to listen to through the speakers of the computer, especially if there is a thunderstorm in the distance, as it picks up the lightning discharges very well. At other times it is best to mute the volume, as it can sometimes sound like an insect trapped in a tin can.

Interpreting the Data

Now that the SuperSID monitor's operating software has been programmed and each station is producing a sunrise and a sunset effect, it is time to start looking for possible solar flares.

All that is needed is a little cooperation from the Sun itself, waiting for the Sun to send out a flare of suitable size, in the right direction (that of the Earth) and more importantly, at the right time of day. It can be quite frustrating, as it may seem like the flares have a habit of happening just before local sunrise or just after local sunset. It's a great feeling when all the waiting pays off and a flare is recorded. In the meantime, while waiting for a flare, it is a good idea to check the graphs each day and familiarize yourself with any unusual spikes and peaks that may be found on the graphs. There are a number of different things that can affect the ionosphere and produce spikes or peaks on the graphs, including lightning, interference, or work being carried out on the transmitter itself. Lightning usually shows itself as a straight vertical line on the graph. Interference can take many different forms and can be of short duration or something more constant.

There is a characteristic shape to solar flares. They have a straight vertical line with a shallow, angled line leading to background levels, somewhat like a shark's dorsal fin. Soon enough, they should be easy to spot.

Look at Fig. 10.13 and the two circled flares. The rise in the graph that starts at about 1700 h is the Sun setting and the ionization in the ionosphere being lost. The two flares rise vertically from the graph, although it is perfectly possible to have the same flares going down from the line. Just by sheer coincidence, two of the stations being monitored are GBZ at 19.6 KHz and the GQD at 22.1 KHz. They have the opposite effect to each other on the graph, as one is a SID and the other is a SES. Looking at Fig. 10.14, this effect can be seen.

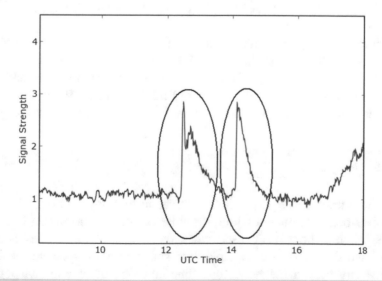

Fig. 10.13 Image showing two solar flares circled

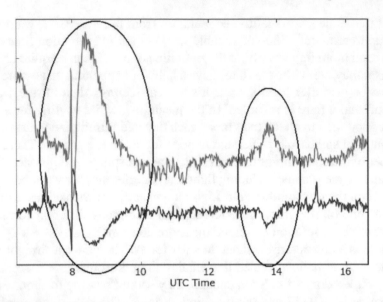

Fig. 10.14 Image showing opposite responses to the signals received by solar flares

The large flare on the left-hand side was an X3-class flare and is quite unmistakable. This flare left the ionosphere quite heavily ionized for some time. The flare on the right-hand side was a C2-class flare. Any suspected flares within the SuperSID can be checked against the 3-day X-ray graph at https://www.swpc.noaa.gov/products/goes-x-ray-flux. This graph is updated every few minutes. Please see Fig. 10.15. The two flares that are circled are the same flares shown in Fig. 10.14.

Looking at the graph, the X-class and C-class flares are both circled. There is a small trace on the SuperSID graph hinting at the next C-class flare. If the NOAA X-ray graph is consulted on a daily basis and then checked against the SuperSID graphs to find the smallest flare your equipment can detect, you can better determine your equipment's sensitivity.

X-Ray Classification

Solar flares are massive explosions, equivalent to many millions of tons of explosives being detonated within the solar corona. As discussed in earlier in this chapter, these are caused when the magnetic force lines become twisted and distorted. For example, two sunspots with opposite magnetic polarity may have a magnetic force line coming out of one sunspot and entering through the other sunspot. Because the Sun is gaseous and not solid, it rotates at different rates at different latitudes. This has the effect of

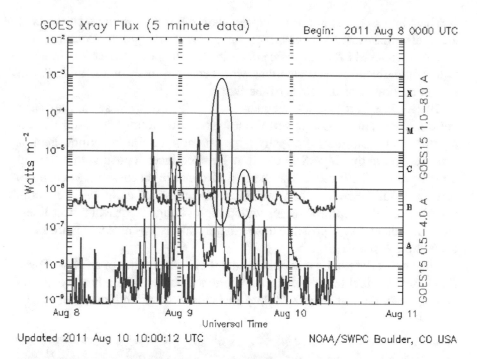

GOES Xray Flux (5 minute data) Begin: 2011 Aug 8 0000 UTC

Updated 2011 Aug 10 10:00:12 UTC   NOAA/SWPC Boulder, CO USA

Fig. 10.15 Image from the NOAA X-ray flux website

winding up these force lines until something has to give. It is like over-stretching an elastic band—the force lines snap and a huge amount of energy is released. This is most apparent during the solar maximum, when the magnetic poles of the Sun are midway between flips. The rotation of the Sun has the effect of winding up the magnetic field lines, like the water coming out of a rotating lawn sprinkler.

X-ray flares are classified into five categories, A, B, C, M and X, based on increasing magnitude and intensity. Each of these categories—apart from the X class, which has no upper limit—are further subdivided into nine different numerical ranges, i.e. A1–A9, B1–B9, C1–C9, M1–M9, and then X1–X<. The A-class flares are the smallest and happen the most frequently. These flares are too small to be detected by the SuperSID monitor. The B-class flares are also too small to be detected by the SuperSID monitor.

The next are the C-class flares, and these are the start of the SuperSID monitors range. The smallest flares detectable on a good day are the C1, but more often the smallest will be the C2 flare. These lower intensity flares can be a little difficult to spot at first, but as the intensity increases to C5 they can get quite large. If the antenna is used at a noisy site, it may be harder to detect the smaller flares.

The M-class flares are less frequent than the three previous classes and are quite unmistakable when seen on a SuperSID graph. Warnings are issued when an M5 flare is detected, as this has the power to cause problems with communication and satellites with sensitive electronics, and also to trigger auroras near the poles of the Earth.

The X-class flares are less frequent than all the other classes of flares and are the biggest and most powerful ones. These are totally unmistakable on a graph, and sometimes if a really large one hits the Earth's atmosphere, it can overwhelm the SuperSID monitor. This produces a peak with a flat top on the graph. These can cause the sort of damage discussed earlier in the section about space weather, including disruption of cellphone networks and GPS satellites, power surges within high voltage electrical power lines, etc. They also make it possible to see an auroral display at lower latitudes— even at the equator.

The largest flare to be recorded to date was in the region of an X28! These flares can be lethal to astronauts, as mentioned earlier. There have been several of these; the last large X28 flare was recorded on 4th November 2003.

Where to Get the SuperSID Monitor

The SuperSID monitors are sold in collaboration with SARA, the Society of Amateur Radio Astronomers, and the Stanford Solar Center. If purchasing a SuperSID monitor for use as an educational tool in a school or university rather than for private use, it may be possible to apply for concessions to obtain the monitor at a reduced cost. By visiting either website, a link can be found to find out how to order a SuperSID monitor. A simple online order form will need to be filled in. The usual delivery information will be needed, plus a site name for the monitor will need to be brainstormed. This can be anything between three and six letters and will then be used with the monitor's personal ID number to process the information if choosing to upload SuperSID site data back to the Stanford server.

The SARA website is: http://radio-astronomy.org.

The Stanford solar centre website is: http://solar-center.stanford.edu/solar-weather.

Other Helpful Information

There is an online help forum where other SuperSID users from all around the world discuss their equipment, ideas, antenna designs, observations and more. Just about anything concerning the SuperSID monitor can be found in this friendly forum, and questions can be posted on the forum pages. Within a short time, quite a few helpful answers with possible solutions will be received.

Visit the Stanford Solar Center website at: http://solar-center.stanford. edu/solar-weather to find out about the online forum and everything to do with space weather. The website also contains lists of useful books and articles about the Sun, and links to other interesting websites. Examples of SuperSID antennas and a list of VLF stations can also be found there.

Other useful websites include:

- http://spaceweather.com
- http://www.exploration.edu/spaceweather
- http://www.window.ucar.edu/spaceweather
- http://www.solarstorms.org/
- http://sohowww.nascom.nasa.gov/

Here is a short podcast about the SuperSID monitor: https://cosmoquest. org/x/365daysofastronomy/2010/04/21/ april-21st-sudden-ionospheric-disturbances/.

The podcast written and recorded by Jim Stratigos, who has made a cracking job of condensing the subject into a 15-min podcast. He discusses matters with a member of the SuperSID team, including about how the SuperSID monitor came into being and how students have found it a great introduction to the subject of studying the Sun. The podcast is a great source of information and well worth the time.

The SARA website can be found at http://radio-astronomy.org. This website has plenty of useful information about the SuperSID and other projects. It is possible to download past copies of their electronic magazine from here, and visitors can read about what other radio astronomers around the world are doing.

Chapter 11

The NASA INSPIRE Project

What Is the INSPIRE Project?

The *Interactive NASA Space Physics Ionosphere Radio Experiments* project, or INSPIRE for short, was created in the early 1990s. An experiment was carried out using an orbiting space shuttle to transmit VLF signals towards the Earth and see if these signals could make it through the Earth's atmosphere, and more importantly the ionosphere, and be detected from the Earth's surface. NASA distributed a large number of INSPIRE receivers to schools all over the United States, as covering a larger area gave the experiment the best chance of success. Thus was born the INSPIRE project.

The INSPIRE receiver is designed to receive frequencies from 0 Hz to 10 KHz. This frequency range covers the designated frequency bands of VLF (very low frequency), ULF (ultralow frequency), SLF (super low frequency), ELF (extremely low frequency) and into the Sub-Hz (sub-hertz) range. At these frequencies, all manner of strange sounds will be heard, both of natural and manmade sources.

The receiver is supplied in kit form and needs to be assembled by the user. The kit contains approximately 60 electronic components such as resistors, capacities, diodes, etc., plus all the sundries such as a box, battery housing and assorted nuts and bolts. The 18-page, A5-sized assembly instruction manual supplied with the kit is more than adequate to build the

© The Editor(s) (if applicable) and The Author(s), under exclusive license to Springer Nature Switzerland AG 2021
S. Arnold, *Radio and Radar Astronomy Projects for Beginners*, The Patrick Moore Practical Astronomy Series, https://doi.org/10.1007/978-3-030-54906-0_11

Fig. 11.1 The INSPIRE receiver as it arrives after unpacking

receiver, although the supplied manual had the odd image missing. But this wasn't a problem, as a complete A4-sized assembly instruction manual can be downloaded and printed from the official website http://theinspirepro-ject.org. If unsure about building the receiver, download the manual and study it prior to ordering the kit. Being bigger, the A4 manual is also handy to have when building the receiver, as the print is easier to read.

As mentioned earlier, if the art of soldering and component identification can be mastered, there should be no problems in building this receiver. The finished receiver doesn't require "tuning," unlike the Radio Jove receiver, and the only test that needs to be performed is to switch it on and listen. This kit is an excellent starting point on the road to building radio receivers, as there are only a small number of parts to contend with and the instructions are very clear and easy to understand. The project when built is a highly portable receiver that can be taken anywhere suitable for receiving VLF emissions. It is battery operated, so there's no need for an external power supply, although there is a facility to plug one in if the need arises. A good quality alkaline battery seems to last quite a while.

The kit can be purchased from the INSPIRE project website at http://theinspireproject.org, or links can be found from the SARA website at www.radio-astronomy.org.

An antenna and a ground stake will also need to be purchased or fabricated in order for the receiver to function, but this will be covered later in the chapter. A pair of headphones and a 9-V pp3 battery to power the receiver will also be needed. After ordering the kit, it arrived in about 2 weeks. Figure 11.1 shows the kit after unpacking.

Note that each plastic bag is numbered. Each one contains all the same components, such as resistors and capacitors, and just needs to be sorted out to identify the value of each component.

A Guide to Building an INSPIRE Receiver

Before starting to build the INSPIRE receiver, ensure that plenty of time is available and try to work when there is little chance of being disturbed. Building the receiver isn't a race. The first thing to do is check that everything is present and correct. The assembly instructions have a list of the kit's contents, as follows, a black plastic enclosure (box), an aluminum face plate and a circuit board. It is very important to handle the circuit board by its edges, like handling a DVD or CD, to prevent the oils in yours fingers from coming into contact with the connections. This oil can be slightly acidic and can cause the solder not to adhere correctly to the connection. This then causes what is known as a dry joint. Notice also that two components have already been fitted to the circuit board. This has been done to save any confusion at the building stage.

Next, check the bag that contains all the sundries, items such as switches, knobs, antenna terminals and other miscellaneous items.

Before starting on the bags that contain the electrical components, try and get hold of some small, plain stickers—something about the size of a price tag will be fine. These can be purchased from stationers or other office suppliers. Mark the stickers R1, R2 and so on, until there are 28 of them. Now open the bag of resistors and lay out the contents. Pick up the first resistor (it doesn't matter which value it is). Using the color code—either the one within the instruction manual or the one within this book—make a positive identification of the resistor's value. Once this has been done, check this value against the list within the instruction manual. At the side of the value, there will be a number from R1 to R28. Stick the appropriately numbered sticker to the wire of the resistor in such a way that it can easily be identified. Take extra care when identifying R3 and R4, as R3 has a value of 2.2 Meg Ω and R4 has a value of 22 Meg Ω, so these can be easily mixed up. Carry on until all the resistors have been identified.

Next, start on the capacitors, but this time marking the small stickers C1, C2 and so on. Some components, like electrolytic capacitors, are polarity-sensitive and *must* be fitted in the correct orientation; all the parts that are polarity-sensitive are indicated within the assembly manual. Do this for all the other items until every one of the components has been positively identified.

This advice will be repeated when describing how to build the Radio Jove receiver. No apology is made for this, as this method may sound time-consuming but has proved to work without fail over and over again, and will save time in the long run.

Next, take note of the positioning of the components on the circuit board. Unlike the Radio Jove receiver, where all the components fit on the same side of the circuit board, the INSPIRE circuit board uses both sides. The resistors, capacitors and other electronic components fit on one side, and the variable resistors, switches and LEDs fit on the other side. The best way to avoid this mix up is to place all of the parts that fit on the front of the circuit board into a separate bag, thus remembering to turn the circuit board over before fitting these components. This may sound obvious, but mistakes do happen.

Some components, like resistors and capacitors, will need to have their connections bent in order to fit them into the circuit board. This isn't as simple as it sounds. Resistors, and especially capacitors, can be easily damaged; if their leads are bent too close to the ceramic body of the resistor or capacitor, the component can fail. This may not happen straightaway but could occur over a period of time, and these failed components can take some tracking down in order to replace them. A good tip to remember when bending the connecting wires is to leave approximately 3 mm (0.125 in.) on either side of the component's body. This can be easily done by using a pair of needle-nose pliers.

Holding the connecting wire with the needle-nose pliers as close as possible to the component's body, bend the connecting wire to the correct angle with your fingers. The pliers will help support the component's ceramic body and stop it from being damaged. Once one side has been bent to the correct angle, place the component in one of the pre-drilled holes on the circuit board. Then the other side can be marked with a marker pen to see where to bend the other connecting wire. Once familiar with the layout of the circuit board and where each component needs to be fitted, you are ready to start soldering each component in place.

The best components to start with are the resistors, as these are less bothered by the heat of soldering, and also resistors are not polarity-sensitive and can be fitted in any orientation.

Find the position of the first resistor, R1. Bend the connecting wires and solder it in place. Then, using a sharp pair of wire cutters, cut off the excess wire after soldering. It's a good idea to have a small container handy to drop these excess bits of wire into, as they can be a problem if they are dropped on the floor and stood on by humans or pets, as a sharp bit of wire sticking into your foot can be very painful.

When all the resistors have been soldered in place, solder the two IC sockets onto the circuit board. Keep the ICs themselves in a safe place until later, as ICs can be easily damaged and should be one of the last components to be fitted onto the circuit board. It is very important that the sockets are soldered in the correct orientation so that the ICs can be correctly fitted later in the process. Look at the socket and find the small notch in one side. This notch denotes the top of the IC that will be fitted later. Look closely at the circuit board to see the corresponding mark where the IC socket is to be fitted. These marks should be lined up together to make fitting the ICs easier later.

Next, solder the capacitors in place. This should be done in the same way as the resistors, starting with C1 and steadily working one by one through them all. Some of the capacitors are polarity-sensitive, and these *must* be fitted in the correct orientation for them to work correctly.

Be very careful with the ceramic-bodied capacitors, as these are fragile and can be easily damaged when bending the connecting wires to allow them to be fitted onto the circuit board.

The electrolytic capacitors are the ones that have a metal casing that looks like a small can. When bending the connecting wires on this type of capacitor, try not to let the connecting wires touch the metal casing, as this could cause a short circuit. The metal casing should be insulated, but it is just not worth the risk.

Next, fit the diodes. Remember from Chap. 5 that a diode is a semiconductor that allows the flow of electrons in one direction only, therefore these are most definitely polarity-sensitive. Looking closely at the body of a diode, you will notice that the manufacturer has marked the diode with a solid color band all the way around one end of the diode's body (the color of the band is not important). Holding the diode horizontally in your right hand with the color band also on the right means that the current flow would be from left to right. The circuit board is marked accordingly, so all that needs to be done is to match up the color bands with those marked on the circuit board. There are zener diodes also included within the kit, and these must not be mixed up with the other diodes. Within the construction manual there are images showing the different types of diodes, and it is very easy to identify which is which.

Now, the two inductors L1 and L2 need to be fitted. These are not polarity-sensitive and their positions are clearly marked on the circuit board. To fit these, very little bending of the connecting wires are needed.

The battery holder needs to be fitted now. This is just a matter of applying four nuts and bolts to hold the assembly to the circuit board and soldering the battery leads to the circuit board to apply power to the circuit.

The positive and negative connections are clearly marked on the board. Please note that there are four nylon washers that need to be fitted between the spacers and the nuts. These are important for later, when the aluminum face plate is fitted.

Working from the other side of the circuit board, it is time to fit the switches, etc., that were placed in a separate bag earlier. Start by fitting the two variable resistors (potentiometers) R7 and R26. The circuit board is very clearly marked for these next few components. The connections for R7 and R26 should be pushed all the way through the pre-drilled holes in the circuit board until the back of the casing is resting on the circuit board. They will only fit one way, so there is no worry about fitting them incorrectly.

Sometimes the connections are overly long, and a little of the connection can be trimmed off before soldering them in place from the reverse side.

The switches should now be fitted. There are five of them with six connections each.

Working one switch at a time, push each switch connection through its pre-drilled holes in the circuit board. Push the switches firmly until the casing of the switch is resting on the circuit board, and solder them in position. The same applies to the switches as R7 and R26—they will only fit one way.

Once all the switches have been fitted, the two LED's should be soldered in place. LED's, being diodes that just happen to glow when electricity is passed through them, are polarity sensitive and must be fitted in the correct orientation. Looking closely at an LED, you will find a rim or collar running around the base of the bulb. One side of the rim will have a flat section that denotes the negative connecting wire. This is shown on the circuit board, so just orientate the LED so the two flats—the one on the LED and the one on the circuit board—line up.

Next, the ICs can be fitted into their sockets. This must be done with great care, as they are easily damaged. The first thing to do is sort out which IC is going into which socket. Since it is just the two of them, this should be quite straightforward. Next, orient the IC so it can be fitted the correct way around; this was why it was so important to fit the IC sockets correctly earlier. Looking at the IC, there is a mark near one end. This mark can be a notch in the centre of one of the ends, or in some cases a colored spot is used to denote pin number one. If the IC is turned so this mark is at the top left— or in the case where there is a notch in one end, this notch is at the top—the IC is now in the correct orientation. Looking at the IC socket soldered onto the circuit board, if the circuit board is then turned until the notch in the IC socket is at the top, or 12 o'clock position, pin number one should now be at top left. If the IC itself is then turned so its notch is in line with the notch on the IC socket, then the IC will be in the correct orientation and can be fitted into its socket.

Once you are sure of the orientation of the IC to its socket, it is now time to fit it in place. Check before fitting that none of the connecting pins are bent. If they are, carefully straighten them with a pair of needle-nose pliers. Take note if the connecting pins are spread too far apart for them to fit in the socket. If they are, carefully holding the IC between finger and thumb, apply slight pressure to one side in order to bend the connections in a little. This is best done by using the surface of a desk or a table in order to apply equal pressure to each connection and hopefully move them all by the same amount.

Offer the IC to its socket again and repeat this operation until the correct spacing is reached. Once the spacing is right, the IC can be carefully fitted into its socket. Fitting an IC for the first time can be a heart-in-the-mouth moment, but by just following a few simple precautions, it can be easily done. Care must be taken to make sure that only one pin enters the connection within the socket. This may seem like stating the obvious, but mistakes do happen, and this is why it is important that the connections on the IC are not bent. It is also useful to work the IC into its socket a little at a time in a slight rocking motion, rather than applying all the pressure to the centre of the IC itself, as in doing this the IC is less inclined to split open.

The next job is to fit a number of short lengths of wire to the circuit board to allow connections to be made to the data and audio outputs on the aluminum face plate.

There are connections for an external power supply, the antenna and ground stake. These connections are sometimes referred to by electricians as *jumpers*. All the wire needed to make these is included within the kit and color-coded to make the job of assembling it easier. The length at which to cut each wire is clearly shown within the assembly instructions. Stranded wire is used, so cut approximately 3 mm (0.125 in.) of insulation off each end of the wire, being careful not to damage the strands beneath. Then, twist the strands together at each end and apply a little solder to stop the strands from untwisting themselves. This also aids in soldering the ends of the wire to the circuit board and the other outputs.

Once each wire has been soldered onto the circuit board, it is time to fit the output plugs, etc. to the aluminum face plate. This is a simple job of just passing the plug through a pre-drilled hole in the face plate and securing it with a locking washer and a nut. These nuts only need a "nip" with a pair of pliers to hold them in place. Please note that within the assembly manual, some of these output plugs are fitted at a slight angle as opposed to being straight; this is to allow them to be fitted into the box later.

Once the input and output sockets have been fitted, the face plate can now be fitted to the front of the circuit board. This is done by removing the nuts and locking washers on the two variable resistors, then placing the face plate in position and refitting the locking washers, tightening the two nuts and fitting two screws into the spaces of the battery compartment fitted earlier. Now, solder the last few connections to join the input and output sockets to the circuit board. Also, solder the connection for the antenna and ground stake that will be fitted later.

The final item to be fitted is the knobs on the front of the face plate that operate the variable resistors to alter the data and audio levels. It's a good idea before fitting these knobs to the shafts of the variable resistors to turn both shafts as far as they will go counterclockwise first. This will position both reference markings on the knobs at the same point on the scale printed on the face plate.

After this has been done, it should look like Fig. 11.2.

Fig. 11.2 The finished circuit board of the INSPIRE receiver

The next thing to do is to fit the completed circuit board and face plate into the enclosure. This is just a matter of applying four screws, one in each corner of the face plate, to hold both the circuit board and face plate to the enclosure. Figure 11.3 shows the completed INSPIRE receiver.

Fig. 11.3 The completed INSPIRE receiver. As can be seen from this image, the completed receiver is a neat little project that is highly portable

It is now time to fabricate or purchase an antenna in order for the receiver to pick up a signal. Looking at the image of the finished receiver, you will notice that in the top left-hand side there is a BNC fitting for use with an antenna with a BNC fitting. This plug must *not* be used to mount an antenna, since if an antenna is fitted to this plug, the receiver will not work correctly.

The INSPIRE receiver is designed to work with an antenna with a length of between 1 and 3 m (39–117 in.). This can be anything from a stout piece of wire to something more elaborate, like a telescopic whip antenna. For the purpose of testing the receiver, a 1-m (39-in.)-long piece of stout wire will be fine. The wire can be fitted directly to the INSPIRE receiver by a screw connection on the top right-hand side of the receiver's face plate. It is a good idea to fold over the last 13 mm (0.5 in.) of the other end of the wire and apply two or three turns of electrical tape to cover any sharp ends in order to stop any potential eye injuries if someone should walk into the end.

A ground stake is needed because this type of receiver needs a reference point, and the best reference point for this receiver is the planet Earth. But for the purpose of testing, it is possible to use oneself, by touching the aluminum face plate of the receiver with a finger. This will do for testing the receiver, but the receiver will work more effectively if fitted with a good ground stake.

A ground stake can be fabricated from almost anything that is made of metal and can be driven into the ground. Some ideas: a long nail, 150 mm (6 in.) or longer; an old metal tent peg; even a corkscrew stake of the sort used to tether dogs. The ground stake must be able to have a connection fitted to it, although this need not be permanently fixed in place. The connection can be made using a crocodile clip on the end of a piece of wire that can be clipped onto the ground stake. It is also possible to use a vehicle's bodywork as a ground point, as long as the engine is *not* running.

Looking at Fig. 11.4, the ground stake is on the left-hand side. It is approximately 300 mm (12 in.) in length and made of copper. A handle was fashioned from a short length of 22 mm (0.886 in.) copper pipe. The tip of the stake has had a point filed onto it to help when the stake is pushed into the ground. The top of an old felt-tip pen stops the point from becoming a danger when not in the ground.

Note the antenna at the top of the receiver. This is a ten-section whip antenna purchased quite cheaply from a local electronic supplier. It measures approximately 1.52 m (60 in.) in length and works very well with this receiver. Alternatively, an antenna removed from an old radio could be used. The antenna could be made as mentioned above from a length of copper wire, but a whip antenna takes up far less room, being telescopic, and therefore is far easier to transport. The antenna is fitted onto a short length of 32 mm (1.25 in.) plastic waste pipe, the type used to take waste water from a washing sink to the drain. This plastic waste pipe is cheap, non-corroding and doesn't need painting. It can easily be purchased from almost any DIY/hardware store. It is a good idea to buy a couple of end stops at the same time, as these can be fitted to each end of the pipe and can easily be drilled to accept the screw fittings for the antenna. The other end of the waste pipe here has had a ¼-in. Whitworth screw fixing fitted so that the whole unit can be mounted onto a tripod. The plastic pipe is fitted to the back of the receiver using two spring clips similar to the ones that may be seen in a workshop holding hammers and other tools to a board.

Figure 11.4 is for guidance only. Others will have their own ideas for the design. Different designs can be found by visiting the INSPIRE website. One interesting design is the "walking stick." This design has the receiver

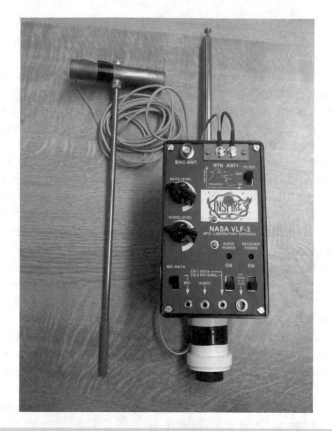

Fig. 11.4 The INSPIRE receiver showing antenna and copper ground stake on the left

antenna and ground stake all mounted onto a piece of 50 × 50 mm (2 × 2 in.) timber; all you need to do is place it into the ground and it is ready to use. This receiver can also be used with a wire antenna that has been suspended between two supports; a simple washing line with a wire core could even be used and suspended between two trees. The differences in designs are only limited by your imagination.

Due to the nature of VLF signals and the problems from interference, it may be necessary to travel to a radio-quiet observing site. Figure 11.4's compact design came into being because of this problem of interference. Everything, including the tripod, is designed to be fitted into a backpack and easily carried to a quiet observing site to be quickly assembled once there. This compact design means that it is also ideal for taking on camping holidays or day trips to the countryside.

After a makeshift antenna and ground stake have been fabricated, it's time to test the finished receiver. A pair of headphones will be needed along with a battery to power the receiver. Before switching on the receiver, turn the audio level knob on the front panel as far counterclockwise as possible. This will save your ears from the possibility of a very loud noise. Next, plug the headphones into the audio output socket, the second from the left. Then, switch on the receiver power switch. On the right-hand side, an LED will light up but no sound should be heard at the moment.

Now, if the audio power switch is operated, the second LED should light up, and it should be possible to hear a noise coming from the headphones. Slowly turn up the audio volume on the receiver until an awful buzzing noise is heard. This buzzing noise is coming from the main electrical wiring within the home and is called *mains hum*. The frequency of this humming noise is the same frequency at which the main AC electrical power is supplied to the home. This frequency can change from country to country. For example, AC current alternates of 60 Hz are used in the United States and 50 Hz in the United Kingdom. So the pitch at which mains hum will be heard is slightly higher in the United States.

Once mains hum has been heard, if the receiver is then moved closer to a power point, the humming gets louder without altering the volume control. Holding the receiver at the side of other electrical devices such as television sets and lights, it will be possible to hear other noises mixed in with the mains hum. Low-voltage desk lamps and other devices that contain transformers seem to be particularly noisy; if these noises are heard it means that the receiver has been built correctly and is working fine.

Notice that the receiver is fitted with a built-in filter operated by a switch on the top right of the receiver. The job of this filter is to help reduce the interference problem by narrowing the window of frequencies that the INSPIRE receiver can receive. If it is found necessary to use this filter, by all means use it, but experience has shown this filter is best left switched off, as the action of the filter reduces the operation of the receiver by too great a margin. It is best to treat this filter like a light pollution filter used on an optical telescope—it is alright as a compromise, but it doesn't beat a truly dark sky. Therefore it is better to try and find a quieter observing site away from interference and radio noise than to use this filter.

This concludes the building and testing of the INSPIRE receiver. The question that arises now is: what type of signals should we expect to hear from the receiver?

Artificial VLF Radio Emissions and How to Choose an Observing Site

The biggest problem in relation to VLF reception is that of interference, as this can come from so many different sources. The worst source by far is that of mains hum, be it 50 Hz or 60 Hz. As nearly every home has some sort of mains/grid power, it can be a little difficult to escape.

This frequency seems to be very penetrating, and if it could be seen, it could be likened to that of light pollution given off from a low-pressure sodium streetlight casting its dirty glow over everything at night, making it hard to identify the true color of things such as vehicles, with the overspill turning the sky a dark orange color.

Unlike the optical astronomer, who isn't bothered by how their neighbor uses their electrical power beyond lighting, the VLF radio observer is interested in every electrical appliance because of the interference that can come from it. Devices with transformers, such as low-power desk lamps, television sets, florescent lights and dimmer switches, are particularly bad, while others, like the normal incandescent lightbulbs, are not.

If living near an industrial area, there will be a battle against the noise from electric motors. Welding equipment is also bad, even if the equipment is fitted with a suppresser. High-power switching gear can also be an issue. Overhead power lines can contribute other problems—if the insulators on the pylons that carry the overhead power lines are in poor order, this can cause interference to be generated around the insulators themselves.

Sometimes there are signals that are superimposed within the frequency of the high voltage electricity. These signals can be instructions for smart meters to switch from a high tariff to a lower tariff and vice versa, and they can produce intermittent problems for radio receivers. Underground power cables are better due to the fact that their interference is somewhat contained within the ground. It has been found that their ability to cause interference depends on a number of different factors, such as the type of ground that the cable is buried in and the depth at which it lies below the surface. If everything has been taken into consideration and as much distance has been placed between the VLF receiver and the nearest main power source, but the receiver is still receiving mains hum, then it is quite possible that there is an underground cable somewhere in the vicinity.

Mains power hum has another annoying problem that a VLF observer should be aware of. It can manifest itself at other frequencies and is called *harmonics*. A harmonic is a frequency that is an integral multiple of the original frequency. For example, if mains frequency is 60 Hz, then it is possible to receive a signal or harmonic at multiples of 60 from the original 60 Hz source—at frequencies such as 120 Hz, 180 Hz and so on. If the mains frequency is 50 Hz, then harmonics maybe received in multiples of 50.

Vehicle ignition systems can also be a problem for the VLF radio observer. Luckily, these can be more of an annoyance rather than a serious problem, because most vehicles will be only passing by. A good idea is to use the INSPIRE receiver at least once in close proximity to a vehicle to learn what to listen out for, as a single-cylinder motorcycle or lawn motor engine can sound quite different to a multi-cylinder car engine.

Clothing is another source of interference of false signals, as was found when testing the finished INSIPIRE receiver for the first time. The receiver is sensitive enough to pick up the very low-level VLF emissions produced by feet on the ground even while walking on grass. Listening intently to the receiver, a strange "whooshing" sound was heard, which turned out to be the low-level VLF emissions generated from one's clothes rubbing together while walking. Analogue wristwatches with a second hand can also be picked up by this sensitive receiver; a pulse can be heard every second as the second hand on the watch moves. Digital display watches don't suffer from this problem.

Cellphones can be a problem; keep them as far away from the receiver as possible. Although not an artificial source of interference, if a flying insect—something along the size of a bee—flies too close to the whip antenna (if this type of antenna is used), then the receiver will pick up the minute electrical signals given off by the insect's wings, producing a buzzing noise in the headphones.

Due to the problem of mains hum, it is more than likely that it will be necessary to travel to get away from this problem. Here are a few important points to consider when choosing a suitable observing site to carry out VLF studies. Try to get at least 800 m (850 yd) away from any form of mains/grid power cables, including overhead power lines, as this will give you the best chance to hear the natural radio emissions and avoid mains hum. This distance should be considered a minimum and if it is possible to get further away, it will be well worth the effort. Avoid valleys that are completely surrounded by hills and mountains in every direction. These will help shield from unwanted interference, but they will also shield the radio signals that we want to receive. Avoid trees, especially very tall ones. The odd tree here and there is unlikely to make a great deal of difference, but don't try to observe from a forest or a clearing in the middle of a forest. Trees contain a lot of water within the trunk, meaning that a wide tree trunk will block the electromagnetic radiation that we want to receive.

Avoid wind turbines. There have been a number of articles indicating that wind turbines can generate strongly within the ELF and VLF frequency ranges, although the articles didn't say exactly how they generate these frequencies. Whether it is from the action of the blades on the air molecules or from the action of the rotor turning within a magnetic field to generate the electricity is not precisely known; it could even be a combination of the two.

Natural VLF Radio Emissions

There are a number of natural radio emissions that are picked up with the INSPIRE receiver. It is probably best to discuss each type and their origin in turn. All these sounds are heard together, but some are more common than others. The first VLF signal that is likely to be heard is *sferics*. The word "sferic" is not a description of the sound that will be heard, but is made from shortening the word "atmospherics." Sferics are of very short duration, typically lasting only a few milliseconds. They can vary in density; some days there are very little, just the odd click every now and then, while other days they are quite noisy and sound like the rustling of a bag of potato chips or the sizzle of bacon frying in a hot pan. When there are a lot of sferics, it can sometimes make it difficult to hear any other VLF signals.

The odd click or hiss can be caused by a random bit of static within the Earth's atmosphere, but the main cause of sferics is lightning. The electrical potential between the Earth and the thunder clouds builds up until something has to give. The air between the thunder clouds and the Earth is usually a good insulator, but any insulator will conduct electricity if there is enough power to force the electrons through the medium. For a brief time the air molecules that usually act as an insulator are overwhelmed and they conduct the lightning bolt. As the lightning bolt strikes, there is a massive release of electrical energy, so much so that sometimes it is possible to smell the aftereffects of the electrical charge in the air. The thunder is generated by the air around the lightning bolt being superheated. This sudden heating of the air makes it expand very rapidly and creates the recognizable noise that is thunder.

In Chap. 4 we discussed the reflective qualities of the ionosphere and how it allows radio waves to "bounce" off it and travel many kilometers from their origin. This is how sferics can travel a few thousand kilometers, by bouncing between the ionosphere and the Earth. If the thunderstorm is close to the observing site, the intensity of the sferics will be greater, and if they become very loud, it may be a sign that the storm is approaching or that it is a particularly bad storm.

The next radio signal noises that we shall discuss are *tweeks*. Tweeks are a short-duration VLF emission lasting in most cases a tenth of a second. The sound of tweeks is hard to describe, but once heard they are quite easy to recognize. Their sound has been likened to a single chirp from a bird, but with a more metallic sound in nature. The origin of tweeks is the same as that for sferics—lightning, but with one difference. In Chap. 10 we discussed the Earth's ionosphere and how it was made up of layers. These different layers are at different altitudes. Tweets are sferics that have traveled up through the Earth's atmosphere and been reflected back towards the Earth's surface from the uppermost layers of the ionosphere. This allows the tweeks to travel tens of thousands of kilometers.

The change in sound between sferics and tweeks is due to *dispersion*. Dispersion is the separation of the radio emissions into its component frequencies. To give an optical example of the principle of dispersion, think of a prism and the effect it has on light. All the optical light wavelengths together make up white light. If white light is passed through a prism, on leaving the prism the white light will have been split up into its component colors or wavelengths. This is due to the different energy levels within in the different colors and produces the familiar rainbow spectrum. In the case of radio wavelengths, our prism is the reflective layer within the ionosphere and its ability to reflect different wavelengths in different ways. This combined with the angle at which the radio waves hit this reflective layer helps split the radio waves into their component parts, as each different frequency will reflect slightly differently. This has the effect of spreading out the radio wavelengths like an optical wavelength spectrum.

The next and most spooky-sounding VLF signals that can be heard are *whistlers*, discussed in Chaps. 1 and 10. There are several different types of whistlers, and they do sound as if someone is whistling.

The first documented evidence of the existence of whistlers was way back in the late 1880s. They were heard through an unamplified telephone earpiece of the time. This telephone earpiece was connected to a telephone line that ran for several kilometers. The whistlers couldn't be heard all the time but only on certain occasions, which must have had the operators of the telephone lines scratching their heads as to what these sounds were and where they were coming from. It was not realized that the long length of the telephone line was acting as a large antenna and collecting the VLF radio signals.

The nature of the whistler and its unusual sound made speculation grow as to its origin. Investigations took place to figure this out, but whatever these whistlers were, the technology wasn't advanced enough to work it out, and some of the interest in these strange sounds faded. It wasn't until the outbreak of World War I when interest in the strange whistling noises came back into vogue. At that time, radio communication was still in the very early stages of development. The equipment needed to make and receive radio transmissions was large and required a lot of power to operate it. Plus, any radio transmissions would probably have been very easy to intercept by the enemy. During World War I, the fighting was conducted from trenches, and a suitable and reliable way was needed for the commanding officers to communicate with the officers in other trenches. So, a network of telephone cables was run from trench to trench, and some of these trenches were large distances apart.

At certain times of the day, but not every day, a strange whistling and hissing noise could be heard coming from each trench telephone earpiece. It was first thought that these strange noises were an attempt by the enemy to block or listen into the communications. Foul play by the enemy was quickly ruled out, and it wasn't until the end of the war when the mystery of these strange noises would start to be understood.

As the technology had yet to be developed, the existence of the sounds relied on the testimony of those who had heard them.

After World War I, tape-recording machines were starting to be developed. These tape-recording machines by today's standards would be classed as rather crude, and the sound quality wouldn't have been great, but it would have been enough to record the strange sounds. The sounds were then processed using a basic spectrogram. A spectrogram plots frequency changes against time (spectrograms will be covered later in this chapter). Using spectrogram analysis, it was finally found that these strange noises were whistlers and other ionospheric phenomena, and that the long telephone lines that ran between the trenches were acting as a large antenna.

It was also discovered that there were different characteristics to some of the whistlers that had been recorded. This gave scientists a clue to their origin. It was found that whistlers originated from bolts of lightning on the Earth, but unlike tweeks, where the radio emission gets trapped between the Earth's surface and the top layers of the ionosphere and bounce between the two, a whistler gets trapped within the Earth's magnetic field. As discussed in Chap. 4, the Earth's magnetic field is produced by a dynamo effect caused by the rotation of the Earth and its inner and outer cores. This magnetic field can stretch a great distance into space. It is thought that the radio emission created by the lightning bolt travels up through the Earth's atmosphere, through the ionosphere and gets trapped within the magnetic field lines of the Earth. It then can be carried along the magnetic field line right into space before being returned back to the Earth with the magnetic field line as it reenters the Earth at the opposite pole. Whistlers can travel great distances around the Earth, and it is not uncommon for them to travel from one hemisphere into space and return to the other side of the Earth in a different hemisphere.

Depending on how whistlers move within the Earth's magnetic field, they can sound slightly different, and with the use of a spectrogram this difference can be seen visually. For greater knowledge of the different types of whistlers, it is highly recommended to read *Whistlers and Related Ionospheric Phenomena* by Robert A Helliwell, as this book covers whistlers in great detail.

The last VLF signal to discuss is the *chorus*. The chorus is a very fitting name for this type of VLF signal it is also fitting for the particular time of day that it can be heard. The best way to describe the sound of a VLF chorus is that moment first thing in the morning, with the dawn chorus of all the birds waking up and chirping. The best time to hear this VLF chorus is also early morning. The VLF chorus is thought to be a result of charged particles contained within the solar wind interacting with the Earth's magnetosphere. These can get trapped within the Earth's magnetic field lines and be funneled down towards the Earth's poles like the aurora. This is why the chorus is best heard from mid to high latitudes in both the northern and southern hemispheres, and not very often on the equator unless solar activity is very high.

Links to Sound Samples

To hear a sample of the different types of emissions that will be received, visit the INSPIRE project website at http://theinspireproject.org/. Once there, a number of examples of artificial and natural VLF radio emissions can be found. It's a good idea to download some sound samples to play from time to time to get familiar with the different types of sounds.

Messages from Beyond the Grave, Myths and Conspiracies About VLF Frequencies

"Have you picked up any ghosts?"

Thanks to modern-day ghost-hunting TV programmes, VLF and ELF frequencies get a lot of press for their use in picking up the paranormal. All the equipment used in these ghost-hunting programmes can be readily bought online, and there are lots of videos on Youtube explaining how to use this equipment.

The INSPIRE receiver should *not* be confused with any paranormal detective equipment on sale, and owning an INSPIRE receiver won't make anyone a Ghostbuster. The INSPIRE receiver is designed to work at such a frequency that it shouldn't contain any commercial broadcasts—"shouldn't" being the word, because if the atmospheric conditions are just right, something that sounds like *but is not* a distorted word may be heard, and the sound of some whistlers can be quite haunting and ghostly when first heard. To show how common this problem is, visit the website www.vlf.it. An enquiry is posted to the website every now and then asking about the possibility of using VLF and ELF frequencies to communicate with lost loved ones.

After practicing radio astronomy for many years, the only supernatural experience witnessed by the author so far occurred while setting up radio meteor detection equipment for a night's recording. When processing the sound file, I heard some very strange noises recorded along with the meteor echoes.

Were these haunting sounds from tormented souls trapped between this world and the next? No! After a little detective work, I found that I forgot to mute the computer's microphone in the audio input settings, so the microphone stayed live all night. Thus, these strange sounds were in fact noises from local visiting hedgehogs and the observatory's squeaking door hinges (these have since been oiled).

Another myth around VL-ELF frequencies is that they can harm living tissue by interfering with the body's natural resonance, such as brainwave activity, and the functioning of other internal organs within the body. Others believe the opposite—followers of the Solfeggio frequencies believe sound played at certain frequencies below 1 KHz can cure anything from the feelings of guilt and fear and repair the DNA of cells within the body. The trouble is that believers of the Solfeggio frequencies can't agree on the frequency! Whether VLF-ELF frequencies do damage or repair health is for science to decide.

Conspiracy theorists have suggested VLF frequencies can be used to communicate with extraterrestrials, but the answer to the question of why extraterrestrials would choose to use these frequencies over any other has not been forthcoming.

Recording VLF

Recording the output from an INSPIRE receiver or any other radio receiver isn't as easy as it may at first seem. The average recording device is designed to record music or speech, and for this reason, most if not all recording devices are fitted with an *automatic gain control* (AGC). An AGC is designed to keep the recording sound levels equal and to stop any sudden changes to recorded input levels. This operation is done automatically by a circuit contained within the recording device itself. This circuit monitors the inputted signal level and either increases or decreases the recording level in order to keep it as constant as possible, without any sudden changes in the sound levels when the recording is played back.

This function is a sort of balancing act. There will be an AGC fitted onto a vehicle's radio in order to maintain a constant volume output from the radio and to stop the volume increasing or decreasing due to any variations in signal strength from the radio's antenna. For example, when driving under a bridge the antenna signal may be reduced temporarily, but instead of the radio's volume being reduced, the AGC will boost the level to maintain the volume at which the radio is set. Once the signal strength has returned after driving out from under the bridge, the gain control will automatically reduce the level again and stop the volume of the radio increasing.

When it comes to recording a signal from a radio receiver like the INSPIRE receiver, a recorder with an AGC would just alter the recorded input so that it would produce a constant level. This negatively affects signals like the tweeks and sferics, since because of their short duration, the AGC would just "clip" these off the recording. AGC's make for good recording of music and speech, but for the radio astronomer, they are nothing but trouble.

Digital recording devices, such as a dictation machine, have excellent recording qualities and can be easily coupled to a computer for quick downloading of the sound file. These also have an AGC fitted so in theory should be of little use to the radio astronomer, but tests have shown they can perform quite well at recording VLF audio signals from the INSPIRE receiver. The one shown in Fig. 11.5 works surprisingly well, and as can be seen, its small size makes it ideal for the job.

If choosing to use a digital recorder, set the sampling rate to its maximum. Another important factor is the format that the sound is recorded and stored. The sound file formats that seem to be the best are WAVE or WMA files. Don't be tempted to use the MP3 sound file format, as the way the

Fig. 11.5 Digital recorder

MP3 sound file is recorded and stored makes it highly unsuitable for use in radio astronomy. The MP3 recordings use a compression algorithm to record the sound, but this compression algorithm removes some of the original sound recordings to save space on the recorder.

The best way to record these types of signals is to use a computer and the sound card that it contains. The quality of the average sound card can in most cases outperform most other digital recording devices. The sensitivity of the computer's sound card in its ability to detect weaker input signals and the frequency responses is far better. The background noise levels of the sound card are far lower than other devices. Operating programs such as Radio-SkyPipe, which will be covered in later chapters, can be programmed to cancel out the remaining noise from the sound card to give a more accurate reading.

To summarize, VLF signals are best recorded on a computer via its sound card, but in cases where it is necessary to travel to a radio-quiet site, a computer may be impractical. In this case, another recording device will be needed if a digital device is to be used. Make sure that the AGC can be switched off (if not, it may still be worth trying). Set the sampling rate to the highest value and only use WAVE or WMA sound file formats to record.

Analyzing Software for VLF

There are a number of good analyzing software programs available that are ideal for analyzing VLF frequencies. The two that will be discussed here are free to download for private use only. Both programs are quite intuitive and easy to use. First is "Spectran," which can be downloaded from www.weak-signals.com. Figure 11.6 shows a screenshot of the program and some of the basic controls.

As seen from Fig. 11.6, the operating program consists of two main windows. The top window gives a graphic representation of the incoming signal, while the window below is a spectrogram image of the same incoming signal. A good way to think of a *spectrogram* is as a graphic representation of any change in frequency of the signal against time. For example, as described previously, the noise from a whistler will be heard to change in pitch during the duration of the whistler. If this change in pitch is plotted against time, it will be shown as a curve on the spectrogram window.

Spectrograms can be useful in identifying unwanted manmade signals. Artificial signals usually have a pattern to them, and this pattern can sometimes be seen within the spectrogram window, for example mains hum or the noise from a large electric motor.

This window can be used in the default setting, whereby the window shows changes in the spectrogram as a waterfall effect, i.e. from top to bottom, or it can be changed to run from left to right if preferred. The color and contrast of the display can be changed to suit personal taste.

In the top left of Fig. 11.6, the main controls are of a slider type. These control volume, speed and gain. Just below these sliders are a series of filters, one of which is very useful. This is the *de-humming filter*, and when this filter box has been checked, it removes most of the mains hum from the signal. For this filter to work correctly, the frequency of the mains hum—either 50 Hz or 60 Hz—must first be set. This can be changed by clicking on the mode button on the top toolbar, then checking the 50 Hz or 60 Hz de-hum box.

If something of interest is seen in either of the two windows, there is a freeze button that will save the screen image to a file so it can be retrieved later. By clicking the "Show controls" button at the bottom right in the image, it is possible to make quick changes to the sampling rates and to make other adjustments to the program even while it is running.

Once the program is started, a "Record" button will appear just below the top toolbar. If this button is pressed, the inputted signal will be saved. A window will pop up asking the operator to input a file name, so a file can

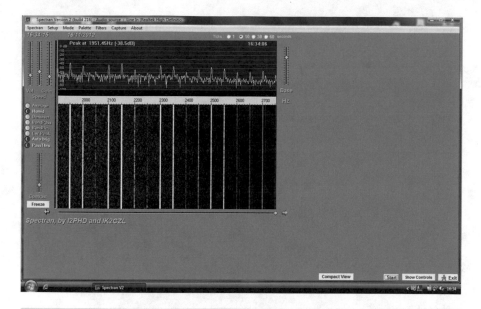

Fig. 11.6 Screenshot of Spectran

be automatically created and then retrieved later. The file format is set by default to the WAVE format. To stop the program recording, it's a simple matter of pressing the "Stop" button.

The controls required to record, play, and pause, are very easy to master. The playback feature of the program allows it to play back recordings made from other devices. This is a great program to experiment with and the settings can be endlessly changed in a quest to find the perfect setting for the purpose. If getting into trouble with the program while experimenting, it has a "Reset" button. If this button is pressed, the program will automatically reset itself back to default settings.

The next program is "Spectrum Lab." This can be downloaded free for private use from: https://spectrum-lab.software.informer.com/versions/ This program is being constantly refined and improved by its creator. It has a vast number of utilities—far too many to mention here, as it would take a whole book to do it justice. The Spectrum Lab instruction manual is well over 100 pages in length. As this program is more advanced than Spectran, only the basic operation of the program is covered here. For those wishing to download this program, it is highly recommend that the latest instruction manual is downloaded at the same time to help with some of the operations of the program.

Figure 11.7 shows a screenshot of the software. At first it will look similar to the screenshot of Spectran. The top window is a graphic representation of

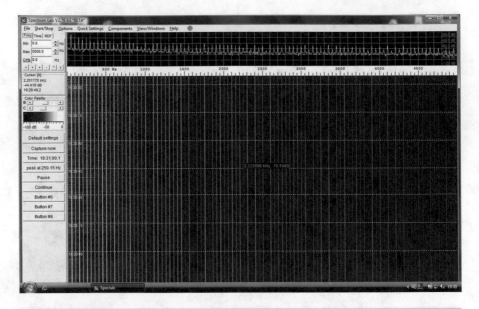

Fig. 11.7 Screenshot of Spectrum Lab

the incoming signal, and the large window below is the spectrogram window. Spectrum Lab has just about everything the amateur radio astronomer could wish to have for processing their data. There are some operations that the program can perform that may never need to be used, depending on the type of signal being processed and how the operator wishes to process them.

There is a very useful feature to Spectrum Lab where it is possible to change everything pertaining to the soundcard and the input mixer on the computer through a control panel window. To change or adjust any particular part of the soundcard or input mixer, just click on the item you wish to change and another window will pop up asking the operator how they want to make the adjustment.

There are numerous options, settings, controls and filters, far more than the Spectran program. It also has the de-humming filter as described above. Even the color of the spectrogram screen can be changed. The program can be used for recording the input from a radio receiver. It will create a sound file automatically and save it so it can be retrieved later for processing in the same sort of way as described with Spectran. It is capable of replaying almost any type of sound file.

There is a "Freeze" button that will automatically save a screenshot of the program, which is useful if anything interesting within the spectrogram window is spotted.

Spectrum Lab also has a "Reset" button, as described above, that will return everything back to default settings.

To summarize: Spectrum Lab is a really useful program, but it can take a while to become familiar with its many operations, although its basic recording and replaying of sound files are quite easy to master. Of the two programs, Spectrum Lab has more useful features, especially if you like to experiment with different filters, etc. But on the downside, it takes longer to master. Spectran is easier to use from the start and is perfectly adequate for the purpose of VLF processing.

Other Helpful Information

Here is the link to the INSPIRE website http://theinspireproject.org/. Once there, it is possible to find all the latest news involving the INSPIRE project, as well as finding examples of different types of antennas, sound files containing samples of all the sounds described above, and more. An INSPIRE receiver can be ordered through this website.

A good source of information that can be found at the INSPIRE project's website is the INSPIRE journal, which publishes all the latest information, useful articles, websites, etc. Some INSPIRE users send in details and images of their receivers and their interesting antenna designs. It is highly recommended to look through some of the back issues of the journal, if only to see how it has changed from its humble beginnings.

Register an email address through the INSPIRE website at http://theinspireproject.org/ to receive updates and other useful information via email.

An excellent website to visit is www.vlf.it. This website is dedicated to the study of very low frequency emissions. It has examples of antennas suitable for the reception of VLF emissions and lots of useful links and applications, such as the use of a processing lab for VLF signals. Note that there may be conditions to it uses. It is possible to pose questions via the site and someone will reply with an answer.

Another useful website this time for software is www.weaksignals.com. This site has some useful software that may be of use to the more advanced radio astronomer enthusiast.

Also consider the book *Radio Nature* by Renato Romero. This is an excellent book for anyone who wishes to pursue the hobby of very low frequency (VLF) observations. It is easy reading with very little mathematics. The INSPIRE project is briefly covered, as is the topic of low frequency radio emissions from the Sun. It has some good illustrations and images, including a rather strange image of a submarine's antenna.

Additionally, there is the book *Whistlers and Related Ionospheric Phenomena* by Robert A. Helliwell. This book explores the topic in-depth and may be a little heavy going at times; it contains lots of spectrogram images, which are very interesting to study, but there is also quite a bit of mathematics involved. Nevertheless, it is a good book if a greater knowledge of the subject of ionospheric phenomena is desired.

Chapter 12

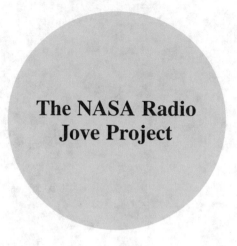

The NASA Radio Jove Project

What Is the Radio Jove Project?

The Radio Jove project is an education and public outreach program involving scientists and teachers from NASA and other organizations. This program supplies a relatively cheap kit capable of receiving radio emissions from the planet Jupiter and the Sun. The kit is supplied in parts numbering a little over 100, and the user assembles the receiver, although a pre-built and pre-tested receiver can be purchased for an extra charge.

The kit as supplied has everything needed to start on the road to running a successful solar and Jovian radio telescope. The only other items required are:

- A 12-V power supply (this can be a battery)
- A pair of headphones or external powered speaker(s)
- And the poles from which to hang the antenna. These poles can be metal, wood or plastic, but plastic has the habit of flexing quite a bit, especially when working at height.

Figure 12.1 shows the kit as it arrives from the Radio Jove team after unpacking. The kit contains five small, numbered plastic bags containing different components, for example resistors, capacitors, etc., along with the circuit board, power cable connections, an aluminum enclosure, antenna connections, six ceramic insulators, a coil of copper wire (from which the

© The Editor(s) (if applicable) and The Author(s), under exclusive license to Springer Nature Switzerland AG 2021
S. Arnold, *Radio and Radar Astronomy Projects for Beginners*, The Patrick Moore Practical Astronomy Series, https://doi.org/10.1007/978-3-030-54906-0_12

Fig. 12.1 The Radio Jove kit as it arrives after unpacked

antenna is to be made), and a coil of RG59/U coax antenna cable along with a very clear and easy-to-understand instruction manual.

With my Radio Jove kit shown in Fig. 12.1, there were two CD-ROMs also included. The CD-ROMs contain lots of useful information and visual aids, plus tutorials on the art of soldering and useful programs such as Radio Jupiter Pro and Radio-SkyPipe. These programs will be discussed later in the chapter. Note that the CD-ROMs are being phased out and are no longer supplied with the kit. Instead, these resources are now free and can be found online, with the exception of the *Listening To Jupiter* book. You will receive a Dropbox link to a PDF copy of that book with your purchase of a Radio JOVE kit. The current links to all the CD information can be found at the end of this chapter.

Before ordering a Radio Jove kit, it is vital to note that the antenna is of a dual di-pole design and will require a reasonably large area to erect it. The two di-poles each measure 7.6 m (25 ft) in length and are placed at a distance of 6 m (20 ft) apart. There are smaller antennas that claim to be able to pick up Jupiter; these will be discussed later in the chapter.

Depending where in the world the receiver is to be used, the height at which the Sun and Jupiter transit above the local horizon will change, meaning the antenna height can range from 3 m (10 ft) up to 6 m (20 ft). If short on space, a single di-pole antenna can be used. This is still able to receive Jupiter, although it will not have the same gain. It can however easily pick up radio emissions from the Sun.

The Radio Jove website is a good place to start, as there is plenty of useful information there to be explored. There is also a list of mentors on the website who can be emailed to ask questions about the Radio Jove receiver. A downloadable electronic newsletter is produced two or three times a year. The newsletters are full of useful information about what is happening in the world of Radio Jove, and they contain images sent in from Radio Jove users all around the world. There is an excellent online server to sign up for, after which you will receive email messages from other Radio Jove users around the world. This server can be of great interest and shows how other Radio Jove users in different countries have received the same radio emission, plus provides any updates to software. Occasionally, they have a phone-in where Radio Jove users can call and discuss all aspects of the Radio Jove project with the mentors.

Building the Radio Jove Receiver

Building the Radio Jove receiver is estimated to take approximately 9 h. This is only an estimate and it may take longer or shorter depending on the builder's capabilities. You will find some similarities to the INSPIRE receiver project; this information is repeated here for convenience of use when building this receiver.

Within the manual there is a list of components along with images of resistors, capacitors, inductors, integrated circuits, transistors, power connections and other items. The manual also features two columns, one for part identification and the other for part installation, each one should be ticked after identification and installation has taken place. If this is done religiously, it will be easy to stop and then pick up wherever you left off.

Before starting to identify the components, it's a good idea to use small stickers— something along the lines of the size used for price tickets in shops. These can be obtained online or from any office supply store. Doing the same as was done with the INSPIRE receiver, work only *one* type of component at a time.

The manual lists capacitors first, and although they number up to 44, there are in fact 43. This is because capacitor number 7 is no longer used but is still listed. After the first capacitor has been positively identified, fold a small sticker around one of the wire connections and mark it C1, then tick the column within the manual. Carry on until all of the capacitors have been identified and labeled. Then do the same with the resistors, of which there are 32. Start with R1 and carry on until all of the others have been identified and tagged. Please note that resistor number 32 does not fit onto the circuit board. Keep this component separate and in a safe place for the moment. Do the same labeling of the other components until each and every one has been positively identified and had the corresponding box in the manual ticked.

Notice that there is a large silver-colored, rectangular-shaped object within the components. This is a 20 MHz crystal and should be soldered in place with the other components on the circuit board. The purpose of this crystal is to tune the receiver, and it can only be fitted in one position so there will be no problems with its orientation.

All this may come over as rather time consuming, but when soldering each part into location, it's just a matter of looking for the right number on the component's sticker, and it also acts as a double check if forgetting to tick the column within the manual. There were two components wherein I found identification to be a problem. These were zener diodes ZD1 and ZD2, as their marking on the case were slightly different than the marks shown within the manual. ZD1 was marked within the manual as 1 N753 but the diode case read F53A, and ZD2 was marked in the manual as 1 N5231 but the diode case read F231B. This is not uncommon when building projects from kits, as different manufacturers have different markings or even different colors for the same component. If this happens to you, a quick search on the internet can help you sort out which is which.

Within the construction manual you will see the term *jumper wires*. These are just short lengths of connecting wire that are soldered onto the circuit board to link different parts together and complete the electrical circuit in a convenient place.

Looking at the aluminum enclosure, don't assemble this yet, but notice the edges are quite sharp from where they have been cut on a sheet metal cutter. Within the kit there should be a small piece of sandpaper. It's a good idea to give each edge a light sanding to remove these sharp edges, which will save fingers from being cut on assembly of the enclosure. After sanding, wipe each piece with a clean, damp cloth to remove any aluminum dust or shavings, as these can cause short circuits if any get in contact with electrical components. These can then be dried and put away for later use, but leave out the front and rear panels.

The front panel has two large holes and one small hole, and the rear has four large holes and one smaller hole. Included within the kit are front and rear decals. These can be peeled off the card and fitted to each panel, taking care to avoid trapping any air, as this can leave unsightly bubbles. After fitting the decals, these panels can be put away with the other parts of the enclosure for later use.

Time to start soldering in the components. First, get the circuit board in the right orientation. Handle it on its edges like a DVD, as oils within the fingers can stop solder from forming a good joint. Look for the Radio Jove 20.1 MHz receiver printed on the circuit board, as this indicates the top and the front edge (see Fig. 12.2). Look at the holes where the components will be fitted; the circuit board is marked with the number of the component that fits in each hole, and the polarity of polarity-sensitive components is indicated.

Components like resistors and ceramic capacitors will need their connecting wires bent so they can be fitted into the circuit board. Don't be tempted just to bend them, as if they are bent too close to their ceramic body this can damage the ceramic cause the part to fail, either immediately or at a later stage. It is better to use a pair of needle nose pliers to hold the connecting wire near the ceramic body to support it and then bend the connecting wire with the fingers. It's also a good idea when bending the connecting wires of an electrolytic capacitor not to let the connecting wires from the

Fig. 12.2 Rear view of finished circuit board (the crystal can be seen at the bottom centre of the image)

capacitor come in contact with the case itself. They should be insulated but it is not worth the risk.

Start by soldering in the resistors first, as they can put up with the heat of soldering. After each component has been soldered into place, turn the board over and trim the excess wire off with a sharp pair of cutters. Have a little container handy to drop these into, as there will be around 200 of these left after building the receiver and they have the habit of getting everywhere.

The last thing to be fitted to the circuit board is the integrated circuits. These can sometimes require a slight bending of the IC contacts so they can be fitted into their IC socket. Be very careful fitting these as they are easily damaged. After finished soldering in all the components, it should look like Fig. 12.2.

Time to fit the antenna input socket, power input socket and audio output sockets to the rear panel. This is just a case of putting them through the pre-drilled holes and applying a locking washer and a nut. These only need slight pressure applied to the nut with a pair of pliers to hold them secure.

The circuit board is now fitted to the base of the enclosure. It is mounted on spacers to stop the bottom of the circuit board from coming in contact with the aluminum base and causing a short circuit. The rest of the enclosure needs to be built now. This is just a simple case of fitting four channels to one of the ends and fixing them in with screws, then sliding in the bottom, front and back panels and securing them with four screws on the other end. The top is left off at this stage, as the connections to the antenna, power supply and audio inputs and outputs need to be made.

The two control knobs for power/volume and tuning can now be fitted. The resistor number R32 (51 Ω) is now temporally soldered onto the antenna input connection. The purpose of this resistor is to simulate an antenna being used with the receiver by producing a "dummy load" on the receiver circuit. This will be removed after testing and tuning of the receiver. Once this has been done, it should look like Fig. 12.3.

The receiver is ready for testing and tuning. This can be done in several different ways depending on the level of equipment available for use. First, a suitable power supply will be needed to power the receiver.

Power Supplies for the Radio Jove Receiver

If unhappy or unsure about using mains/grid voltage, ask the advice of a qualified electrician first.

A 12-V power supply is needed for the Radio Jove receiver. This can be a battery, such as a car battery. Either use the battery on its own or working through the vehicle's cigarette lighter socket, *not* with the engine running.

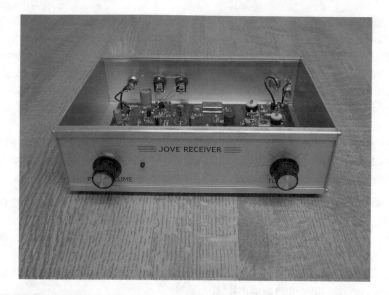

Fig. 12.3 The circuit board fitted into its enclosure after the final connections have been made and the temporary R32 (51 Ω) has been fitted

If mains/grid power is to be used, beware that a well-regulated transformer is needed, not the cheap ones used for running gaming consoles.

Batteries produce power through a chemical reaction and therefore produce true direct current (DC). A transformer works off an alternating current (AC) and steps it down from mains/grid voltage to a smaller voltage, in our case 12 V to run the Radio Jove receiver. Many of the cheaper transformers' outputs have very poor rectification properties. If the output from one of the cheaper transformers is seen on an oscilloscope display, it will show rippling or spikes. These spikes and ripples within the power supply introduce noise within the receiver's electronics. Also steer well clear of the transformers that have a voltage selection switch on them. Experience has shown that the voltage marked on the transformer has nothing whatsoever to do with the voltage coming out! These are alright for running games, etc., but not for the Radio Jove receiver or other radio receivers discussed here.

If the output of a battery or a well-regulated transformer were to be displayed on an oscilloscope, a smooth, straight line is seen. This is ideal.

Testing and Tuning the Receiver

Before switching on the power, make sure the polarity of the power supply is correct. The inner part of the power connection is positive and the outer part is negative. Do a quick check with a multimeter set on the voltage setting. Apply the test probes to the output of the power supply with the positive probe at the centre and the negative on the outer part. The voltage reading should be 12 V. If the reading is −12 V, the polarity is wrong and must be changed before it is applied to the receiver, or the receiver will not work or be permanently damaged. Once the power supply has been correctly identified, it can be plugged into the power input socket of the receiver.

The receiver must be "tuned" before it can be used. There are several ways of doing this depending on the level of test equipment available. What follows is a brief description of each of the ways it can be done.

1. If no test equipment is available to tune the circuit, listen to the tone of the receiver and adjust the tuning until a constant pitch is heard.
2. Use the program Radio-SkyPipe to chart the changes in the output of the receiver. It will be possible to see any drifting on the graph produced by the Radio-SkyPipe program on the computer screen, and changes to the receiver can be made accordingly.
3. Use a multimeter that is capable of measuring audio frequency voltages to measure the output from the receiver itself.
4. Use an oscilloscope to monitor the output. This gives a visual reference to watch to see how the output changes. See Fig. 12.4.

Fig. 12.4 Testing of the Radio Jove receiver using the oscilloscope method and externally powered speakers

A good idea is to have either headphones or externally powered speaker(s) on hand. After choosing the preferred method of tuning, and after all necessary test equipment and headphones or speaker(s) have been fitted, its time to switch on the receiver for the first time. Then apply the power. There should be a high-pitched noise coming from the headphones or external speaker(s). This is the noise being generated within the circuit from the 20 MHz crystal mentioned earlier. Run the finished receiver for about 5 min before starting the tuning process, as this will allow the receiver circuitry time to "warm up" and stabilize, which will lead to a more accurate tuning.

It is now a matter of adjusting four components to tune the receiver. First, set the tuning knob on the receiver to the 10 o'clock position, and using the plastic tools included within the kit, adjust the component L5 (variable inductor) until a loud tone can be heard. Once this is done, forget L5 and move on to components C2 (variable capacitor), C6 (variable capacitor) and L4 (variable inductor). These need to be adjusted until the maximum signal strength is produced. Only make very small adjustments at a time, being careful not to force the adjustment screw in either direction as this can damage the internal parts of the component itself. This may sound a little fiddly, but in fact it is quite simple, especially if using the oscilloscope method, as the actual change in strength of the signal can be seen visually. Not everyone will have access to an oscilloscope, but there is no reason why the other tuning methods shouldn't work just as well.

Once the tuning procedure has been completed, and with the power to the receiver switched off, the resistor R32 (51 Ω) can be unsoldered. This resistor must be kept safe in case this tuning procedure needs to be carried out again. A good idea is to stick it to the inside of one of the panels of the enclosure with electrical tape for future use.

One last thing to do before fitting the top to the enclosure is to cut through the number 6 jumper connection wire, but don't remove it—simply snip through the wire and bend it up so it cannot come in contact with any other part. This removes the tuning crystal from the circuit now that it has completed its job of tuning the receiver, but by not removing the jumper wire, if the tuning process needs to be repeated for whatever reason, it can be more easily reconnected.

With all of the above tasks completed, the receiver is ready for use, as shown in Figs. 12.5 and 12.6. Please note the connections on the rear view of the receiver, in particular the two audio outputs. This is set up in such a way that one output can be used with headphones and the other can be used to connect the receiver to a computer or other device.

Fig. 12.5 The finished Radio Jove receiver front view showing the volume control on the right and tuning control on the left

Fig. 12.6 The finished Radio Jove receiver rear view. Please note the connections (from left to right): antenna input, audio 1, audio 2 and 12-V power input

Calibration of the Radio Jove Receiver

Although the Radio Jove receiver will work without being calibrated and any radio observations are still good, they are not as useful as calibrated observations, as these can be compared with observations made by other Radio Jove users where everyone is using the same unit.

In radio astronomy, the term *antenna temperature* is used. Antenna temperature is a measure of the power per unit of bandwidth of the antenna and is given in degrees Kelvin (K). This can require some serious mathematics to explain; a thorough explanation goes beyond the realms of this book. To keep it simple, the antenna temperature is not a physical measure of the antenna temperature itself, but rather a measure of radio energy the antenna is receiving from a radio source.

For example, the calibrator discussed below simulates an antenna temperature of 25,000 K into the Radio Jove receiver. Think of temperature as not being either hot or cold, but rather as a measure of the energy at which atoms themselves vibrate or move. At absolute zero, or zero degrees Kelvin, all the vibrations and movements within the atoms of a material stop. These atoms have no thermal energy left.

We need to input a signal of a known value in order to calibrate the Radio Jove receiver to the program Radio-SkyPipe. A way to do this is to use the RF-2080 CF (Fig. 12.7). It is supplied ready-built and comes in two different types. The first is the RF-2080 C. This particular model only has the calibration unit fitted and is a good chose if living in a radio-quiet site, but for most of us, the RF-2080 CF will prove a better purchase. This particular model has a narrowband radio filter fitted alongside the calibrator section of the unit. This filter has proven itself useful in blocking some of the radio interference from observing sites. The downside is a small loss of signal strength from the antenna to the receiver due to the internal circuitry of the filter. Yet, the benefits of the filter outweigh this small signal loss.

The calibrator is fitted between the antenna and the Radio Jove receiver. When the calibrator is switched on, it introduces a signal of a known quantity and power (25,000 K) into the Radio Jove receiver. Using the program Radio-SkyPipe, click onto the tools tab at the top of the screen and scroll down the list. The last on the list is the calibration wizard. Selecting this starts the calibration process. There are a number of on-screen instructions, all of which are quite easy to follow. Once done, the program Radio-SkyPipe will automatically carry out the calibration. Unless the volume control on the Radio Jove receiver is altered, the calibration will hold fast. If the volume control is moved, then the calibration wizard will need to be

Fig. 12.7 The RF-2080 CF calibrator/filter

run again. This is why a volume control on the headphones is desirable, so as not to disturb the calibration.

This calibration process will provide you with insight into whether the observing site is radio-quiet or not. As mentioned, the calibrator simulates an antenna temperature of 25,000 K. This is generally considered to be a radio-quiet site. Operate the switch on the calibrator. This will switch off the calibrator and switch back to the antenna itself. If there is not a large increase in noise, then this is a radio-quiet site and there should be no problem in receiving radio emissions from Jupiter or the Sun. If there is a large increase in noise, this indicates a noisy radio site. Don't worry, all is not lost. If a trace on Radio-SkyPipe is started, this will give an indication of the level of background noise at the observing site.

A good idea is to leave the Radio Jove receiver and Radio-SkyPipe running over the course of a few days to detect patterns within the trace itself. Look for times when there are spikes on the graph, likely from lightning. Interference can come from almost any electrical device, but some are worse then others. Interference from such electrical devices are generally reduced from around midnight to 5 am, when everyone has switched off their television sets and gone to bed. It is not uncommon to get the odd spike of interference now and again from heating thermostats, especially on cold nights.

If the observing site has a background noise level constantly above 75,000 K, this means that there will be very little chance of picking up the planet Jupiter. The radio storms from the giant planet don't exceed this level, so in this case, try and find a quieter site for the antenna. Other Radio Jove users have found that moving their antenna a short distance from its original location can make all the difference. If this is not possible, it should still be able to pick up the radio emissions from the Sun, as these can reach levels of several million degrees, although some of the weaker radio emissions from the Sun may be missed.

If lucky enough to have a quiet site, run the Radio Jove receiver over a number of days. It may be noticed that there is a rise in the background levels on the Radio-SkyPipe graphs. This will appear almost at the same time each day, but 4 min earlier each day. This rise in the background level is the same phenomenon Karl Jansky picked up with his merry-go-round antenna in the early 1930s: it is the galactic centre in the constellation of Sagittarius. One lucky Radio Jove user has even picked up a pulsar using their equipment.

Antennas for Receiving Jupiter and the Sun

There are two antennas that can be used to receive radio emissions from the Sun and Jupiter at 20 MHz. Both require a large area of ground to set up. The first is the Yagi antenna discussed in Chap. 6. If the observer has the room, a Yagi antenna will deliver the greatest gain and a more powerful signal. The downside with this type of antenna is its physical size. For example, if a four-element Yagi antenna is to be used, the length of the boom is 5.5 m (18 ft). If a five-element Yagi is to be used, the boom length increases to 7.32 m (24 ft). At these sizes, the antenna would need mounting on an antenna mast. This creates another problem, as a Yagi antenna has a narrow beam width, so the antenna needs to move to keep the object within the antenna beam as the Earth rotates. This can be done manually ever 10–15 min or an electronic drive can be fitted so the antenna can be moved at the touch of a button. Unfortunately, antenna masts and electronic drives don't come cheap.

Most people will not have the room for a mast and a Yagi antenna of this size, so the next best option is to use a dual dipole antenna supplied with the Radio Jove kit. The Radio Jove dual dipole antenna still requires a fair-sized piece of ground to set up. As an absolute minimum, the antenna needs 7.62 m (25 ft) in the east-west plan and 6 m (20 ft) in the north-south plan. The height of the antenna may also prove a challenge, as the minimum

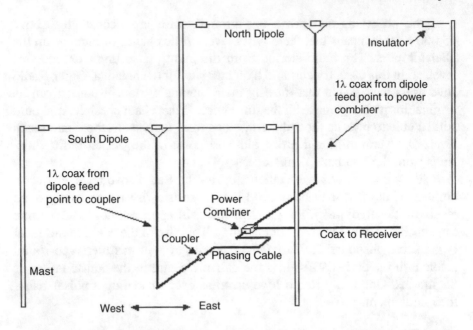

Fig. 12.8 Layout of the Radio Jove dual dipole antenna (Image courtesy of the NASA Radio Jove Project)

height is 3 m (10 ft) rising up to 6 m (20 ft). These measurements don't include room for any guide ropes that may be needed to support the antenna masts. See Fig. 12.8.

The radio Jove dual dipole antenna has a smaller gain but a much wider beam width so the object stays within the beam of the antenna for longer, meaning it doesn't need to be moved. A good thing about this antenna is that it can be taken down when not in use and coiled up for easy storage.

If room is tight, a single dipole can be used to study the Sun. Solar radio emissions are quite strong and a single dipole will pick them up. Picking up radio emissions from Jupiter using a single dipole is possible but requires a radio-quiet area in which to work effectively. Some observers never manage to pick up radio emissions from Jupiter even when using the dual dipole antenna. This is because they live in a radio-noisy environment, but most observers are quite happy just picking up radio emissions from the Sun.

Antennas *Not* to Use for Receiving Jupiter and the Sun

Anyone new to the subject may do a little research on the internet to find out what equipment to use or look for a smaller alternative than the Radio Jove dual dipole antenna. One antenna that will be shown on an internet search is the DDRR (Direct Driven Ring Radiator) antenna. This antenna was invented by Dr. Boyer in the 1950s for the military. It stayed classified until 1963, when Dr. Boyer published an article in a magazine about its design. His antenna was constructed of a large loop made from 100-mm (4-in.) tubing above a solid metal ground plate. The ground plate in this case was the surface of a warship.

Here is a link to an article about Dr. Boyer and the DDRR antenna: http://www.orionmicro.com/ant/ddrr/ddrr1.htm.

In 1989, an article was published in *Sky and Telescope* that described a small loop antenna capable of picking up Jupiter and the Sun. This loop antenna was constructed from a square piece of plywood approximately 762 mm (30 in.) square. The plywood base was covered in chicken wire to act as a ground plate. Protruding from the base were four round dowels approximately 300 mm (12 in.) in length. The dowels had a hole drilled in the end and a wire was threaded through the holes to produce a loop of wire approximately 500 mm (24 in.) in diameter. This antenna was in fact a small DDRR antenna.

There is a group of astronomers in the UK who claim to be able to pick up Jupiter using one of these small loop antennas and a portable shortwave radio. Here is a link to their website: http://www.ukaranet.org.uk/uk_amateurs/bobgreef/.

Anyone thinking of trying one of these small loop antennas may be interested in reading an article published by J. Brown and R. Flagg (the author of the book *Listening to Jupiter*). Here is a link to this article, which can be downloaded as a PDF: http://radiojove.org/SUG/Pubs/Evaluation%20 of%20DDRR%20vs%20JOVE%20Dipoles,%20Brown%20&%20 Flagg%20(2017).pdf.

In the article, J. Brown and R. Flagg directly compare a small loop/DDRR antenna with the Radio Jove dual dipole antenna. They constructed a small loop antenna, carefully following the article first published in *Sky and Telescope* magazine. Both antennas were compared at the same time. The tests were simply to see how each antenna performed at picking up the Sun, the galactic background and Jupiter. The last image in this article sums up the tests results perfectly when the small loop antenna is thrown onto a bonfire!

Constructing the Radio Jove Dual Dipole Antenna

In this section, the construction the Radio Jove dual dipole antenna will be covered, using items supplied with the Radio Jove kit. The dual dipole design gives greater gain than a single dipole and uses a phasing cable. This when used in conjunction with the height of the antenna and the ground effect discussed in earlier chapters has the result of positioning the centre of the beam and allowing the user to get the most gain from the antenna.

The tools needed are:

1. Soldering iron(s)
2. A solder. 1 mm (0.040 in.) diameter is a handy size for most jobs
3. Pliers
4. A sharp craft knife
5. Wire cutters
6. A tape measure

A half-wave dipole antenna is an antenna cut to half the wavelength of the incoming radio wave that is to be received. In this case, at 20.1 MHz, the half-wavelength distance is 7.09 m (23.28 ft).

The six ceramic insulators allow the antenna wire to hang in free space from the mast supports without allowing the antenna wire to come in contact with either of the supports or the antenna feed cable. This is done by fitting an insulator at the end of each of the antenna wires and one in the centre of the wire where the antenna feed cable is connected to the antenna wire itself. A sharp pair of wire cutters will be needed to cut through the antenna coax cable. Nothing comes across as more amateurish than a length of cable that looks like it has been hacked through with a blunt pair of scissors. This can cause all sorts of problems when trying to fit the connections later on.

The first job is to unroll the coil of copper wire and cut it into the correct lengths to make the two dipole antennas. It can be quite awkward trying to measure the correct lengths of wire while it is in one long coil, as it will have the tendency to keep rolling back on itself like a roll of wallpaper. A good way to cut the antenna wire to save on measuring out long lengths is to cut it in half and then half again to have four equal lengths of wire. These will still be too long to make the antenna, but the length allows enough wire to pass through the ceramic insulators and twist it back on itself to secure the wire to the insulators.

When the four equal lengths of wire have been cut, start to fix them to the insulators at the correct length. A good way to do so is to measure out the correct length on a floor and mark this with two chalk marks. This will save and effort trying to hold the antenna wire and stop it coiling back on itself while at the same time trying to hold a tape measure. It helps to have assistance with this part. Hold one insulator on one chalk mark and have an assistant hold one insulator on the other chalk mark. Then get hold of one of the lengths of antenna wire. Each person can allow the same amount of spare wire at each end to allow this to be passed through the insulator. After passing the wire through the insulator, twist it back on itself. Don't go right up to the insulator, since if fitted too tightly, it may cause the ceramic to crack or the wire may become damaged as it rubs against the hard insulator when the antenna moves in the wind.

Once this task has been performed, there should be two antenna wires of the correct length with an insulator at either end and one in the middle. Don't solder these in place just yet. The two antennas can be carefully coiled up and put somewhere safe for the time being. When coiling the antenna wires, try not to let the ceramic insulators knock together as they can soon have their protective glazing damaged.

Here are a few points to follow before cutting and handling the antenna coax cable. The coax cable has a natural twist already formed within the cable itself. When coiling this type of cable, allow it to follow its natural twist, and don't make the loops too small. Doing this allows the cable to be coiled and uncoiled more easily, and the central conducting wire, which is solid in this coax cable, will not become stressed, is less likely to break and also looks neater.

Under no circumstances should you allow antenna wires or the antenna coax cables to become kinked. If needing to take the antenna coax cable around a corner, do so with a gentle curving of the wire rather than a sharp bend. This may sound like laboring the point, but if these simple precautions are followed, there is no reason why the antennas shouldn't provide many years of service.

The antenna cable needs to be cut at set lengths so both of the dipole antennas will work correctly. The lengths of each cable are cut in factors of the wavelength of the signal that the receiver is going to receive (as discussed in Chap. 7). There are four cables to cut. A good idea is to wrap a different color of electrical tape around each of the four lengths and make a note of which color corresponds to which length, as it is hard to judge this when they are coiled up, and this will save time later.

All cables must be cut to their exact measurements. The first is 9.58 m (32.31 ft). There are two of these. Now cut the phasing cable. This is cut to 0.375 of the length of the wavelength, and this equates to 3.69 m (12.12 ft). The last cable to be cut is the cable that comes from the power combiner to the receiver and is 0.5 of a wavelength, which is equal to 4.93 m (16.16 ft). If this is not long enough to reach the receiver, this can be extended in 0.5 wavelengths up to a maximum of 5 wavelengths in length.

Next, get one of the lengths of cable that will be fitted to the dipole antenna wire. Slide three ferrite collars on to the end of the antenna coax cable, which will be fitted to the antenna wire. These improve the antenna performance and help stop interference entering the open end of the coax cable. These can be a tight fit, but they should be easy enough to push on. Try not to knock these together or force them on, as they can be easily damaged. Slide them down around 500 mm (2 ft) and apply a little electrical tape to temporarily hold them in place while soldering the connections.

Now strip the outer sleeve of insulation off the end of the antenna coax cable to a length of 100 mm (4 in.). Take care not to cut into the braided layer underneath. Two or three pieces of braiding will probably be lost. This is normal, as long as it doesn't run into half a dozen or more.

Now carefully untangle this braiding off the inner sleeves insulation. This can be done using the tip of a small screwdriver or the tip of a ballpoint pen, as this has a nice round end with no sharp edges to it. Once this has been done, twist together all the braiding strands to make one wire. This will be one of the connections.

Next, cut 50 mm (2 in.) off the inner insulation sleeve to expose the central conductor. The central conductor on this type of coax cable is a single solid one. Be careful not to score the central conductor with the craft knife as this will cause the wire to have a weak spot, which could be damaged when the antenna is blowing about in the wind.

Making the connections to the antenna wire, first place the antenna coax cable over the central insulator and the antenna wire, holding it in place with a couple of cable ties. See Fig. 12.9.

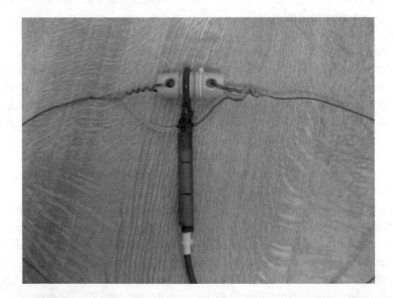

Fig. 12.9 The connection of antenna cable on the central insulator of one of the dipole antennas. Note the three ferrite collars seen on the antenna cable below the central insulator

Once the antenna cable has been secured, the connections need to be made from the antenna feed cable to the antenna wire itself. Start with the braided wire. Twist this around one of the antenna wires on one side of the central insulator. Don't pull the wire too tight; a little slack should be left to allow the antenna to move in the wind. Next, fit the solid central conductor to the antenna wire on the other side of the central insulator. Special care must be taken here. With a gentle curve of the antenna cable's central conductor, bring it up to the antenna wire, allowing enough slack in the antenna wire for it to move without undue pressure being applied to the solid central conductor, for the same reason as above.

Because of the thickness of the wire, a powerful soldering iron will be needed one in the region of 75 W. Don't try and use the 25–30-W soldering irons that are sold in hardware stores, as these will not produce adequate heating and will produce what is known as a dry joint. A *dry joint* is where the solder just sits on the wires and hinders the flow of electrical energy. Enough heat must be applied to allow the solder to flow freely through all the antenna wire connections. It may be possible to hire a 75-W soldering iron for the day from a tool store.

Once the antenna coax cable has been soldered onto the antenna wire, solder the twisted antenna wire on each of the insulators at each end of the antenna wire to stop them from coming free. After the soldering has been completed, slide the ferrite collars back up to the top of the cable and hold them in place with a cable tie, as shown in Fig. 12.9.

The open end of the antenna feed cable must be waterproofed to keep out damp, moisture and dirt. One way of doing this is to apply electrical tape to the open end of the antenna cable and then apply a waterproof coating to this. Use waterproofing specially made for coaxial cables, or an automotive ignition sealer can be used—not the type used to dispel moisture, but the type that leaves behind a thin waterproof skin. Spray a liberal amount all around the open end of the antenna cable. It doesn't matter if the spray covers other parts such as the insulator. Allow this first coat 10–15 min to dry, then apply a second coat of the spray. This should be adequate to waterproof the cable. Depending on annual rainfall, if this task is repeated two or three times a year when the joint is dry, this should be enough to top off the waterproofing action of the spray. If this is done right, a can of sealer will last for years. A word of caution: don't use the spray indoors, as it has a terrible chemical odor to it that will fill every room in the home.

The other dipole antenna can then be constructed in the same way. Once both dipoles have been finished, the F connectors will need to be fitted to the ends of all the antenna cables. Start by removing 25 mm (1 in.) of the outer sleeve of insulation from the antenna cable, being careful not to damage the braiding underneath. This time, only untangle half the length of the braiding and fold this back on itself over the un-braided part. Remove the insulation from the solid central conductor, being careful not to score the central conductor itself.

Now, fit the F connector. This is done by a clockwise twisting motion. Apply firm pressure until the threads on the inside of the connector "bite," and then the connector will screw itself home. Hand pressure should be enough to fit the connector. There should be no need to force it on with pliers, although it can get a little tight towards the end and a small amount of pressure can be applied.

Once all the F connections have been fitted to the antenna cables, wrap a little electrical tape around the end, starting with the tape half on the F connector and half on the antenna feed cable. Two or three turns should be enough. This precaution helps to stop dirt and damp entering through the gap between the connector and the outer sleeve of insulation and prevents corroding of the antenna cable.

It's a good idea to use different colored tape to indicate which cable is which.

Constructing the Antenna Masts

The antennas are used horizontally polarized and need to be mounted at a height that will be suitable to get either the Sun or Jupiter, as close to the centre of the antenna pattern as possible. The height at which the antennas are mounted can be anything between 3 m (10 ft) minimum and 6 m (20 ft) maximum depending on where the antennas are used in the world and the altitude at which the Sun and Jupiter appear in the sky above the observer's local horizon.

Four masts will be needed, from which we will suspend the antennas, along with guide ropes to help steady the mast. The supports can be made of plastic tubing, wood or metal. Plastic is cheap, corrosion-resistant, light-weight, and easily carried in sections to an observing site and assembled there. However, long lengths of plastic tubing can become flexible and prone to splitting or breaking. Choosing plastic tubing will probably limit the antenna height.

Wood is a great all-around choice. It's relatively cheap to purchase but in long lengths can prove a little awkward to handle. If a permanent site is proposed, then wooden masts permanently fixed into the ground and supported with guide ropes would be a good idea. The only maintenance needed would be the application of a timber preservative. If the antenna supports are permanently sited, it would be a good idea to engineer some sort of pulley system that can remotely alter the height of the antennas from the ground.

Metal supports are more expensive than wood or plastic, and they are heavier and prone to corrosion. However, they are reasonably rigid at longer lengths, especially if supported with good guide ropes. One idea is to use metal washing line props, as these are good antenna supports, are available from most hardware stores and are quite cheap. They are telescopic and can be collapsed down like a whip antenna on a car for easy storage and transport. When folded down, they are no longer than about 1 m (39 in.) in length. They are alright for making observations where the antenna is used up to 4.5 m (15 ft). Above this, they are unsuitable, and more rigid supports are needed.

Eight of these will be required. Cut off the plastic hook at the end of four of them. Leave the hook on the other four. This is ideal for hanging the antenna. Next, taking the four poles with the plastic hooks still attached and remove the plastic bung from the bottom of each pole, so that the thinner of the two poles from the one that has had the hook cut off can be slid inside. Two poles can be easily fitted together to form four longer supports. The poles can be drilled to accept fixings for guide ropes. A good idea is to make hoops to fit into the holes in the poles rather than passing the guide rope through the holes in the poles, as any rough edges on the inside of the drilled hole will act as a saw and cut through the guide rope by the motion of the poles moving in the wind. These hoops can be made simply by bending some small-diameter metal into a triangular shape and fitting the two open ends inside the poles. These can be made using old round tent pegs or wire coat hangers.

These poles, if erected properly and equipped with guide ropes and checked from time to time, can cope with quite bad weather conditions. Sometimes these poles come with some kind of plastic protective coating or are even galvanized so no painting is required. The quality of this protective coating isn't great, but they should last a few years before corrosion sets in.

A point worth mentioning about guide ropes. Washing lines are fine, as they are cheap and come in long lengths, but some come with a metal core surrounded by a plastic sleeve or coating. These must be avoided, as the metal within the washing line will disrupt the shape of the antenna pattern, as will any metal nearby, such has a chain-link fence. The above metal antenna poles will not affect the antenna pattern, as they are positioned at the antenna's minimum pickup (see Chap. 6). Avoid ropes made with natural fibers, as these can soon rot if they are constantly getting wet, and they break down when exposed to ultraviolet radiation from the Sun unless they have been treated with an ultraviolet protective coating. The best rope for this purpose is nylon rope, which is lightweight, very strong, and because it is a washing line and therefore designed to be used outside, it will have some kind of ultraviolet coating applied to it. When cutting nylon rope, a good idea is to melt the fibers of the cut ends with a gas cooker ring or candle to prevent the rope from fraying.

Guide ropes can be anchored to the ground using metal, wooden or plastic tent pegs. Avoid the small round type of pegs as these will be forever coming lose and needing to be hammered in again. Use the better quality triangular-shaped ones; these are more expensive but worth it in the long run.

Antenna Configurations

If an observer does not have enough room for both dipoles, they might wish to try using a single dipole to pick up the Sun. The ground effect can still be used to point the beam in the same way as explained below, although the beam width and the gain of a single dipole antenna will be reduced compared with the dual dipole arrangement. Still, it will pick up strong radio emissions from the Sun.

The Radio Jove dual dipole antenna used has many different configurations. These are designed to get the maximum gain out of the antenna when the Sun and Jupiter are at different altitudes in the sky as seen from the observer's local horizon. The configurations described here will be suitable for most mid latitudes in the northern and southern hemisphere. If used in the southern hemisphere, all that is required is to change around the phasing cable so the antenna pattern then becomes north-phasing.

Without the use of the phasing cable, the antenna pattern of the dual dipole antenna would produce two radiation patterns with equal pickup areas that would look like two inner tubes, one for each dipole (as discussed in Chap. 6). By fitting the phasing cable, you can produce a larger gain on the south-facing antenna pattern (north if used in the southern hemisphere).

This is not to scale, but to help picture this, think of one dipole as having the inner tube size of one used on a small car tire. Picture the other dipole with the phasing cable as having an inner tube as large as one used on a truck tire. The large inner tube would be flattened into an elliptical shape, which lowers the altitude at which the maximum gain is achieved. Using this in conjunction with the height at which the antenna is placed, we can capitalize on the ground effect (as explained in Chap. 6) to further refine the shape of the radiation pattern and push or pull the center of the antenna beam where we want to get the most gain.

The following few figures are screenshots from the computer program Radio Jupiter Pro (this software will be covered in more detail later). It will be easier to explain the antenna patterns and the effect of the height at which they are used by referring to these screenshots.

Looking at Fig. 12.10, the antenna pattern is the ellipse shape in the middle. The two horizontal lines in the middle represent the two dipoles, and the cross in the very centre is the point of maximum gain of the antenna in this configuration. Jupiter is seen just inside the antenna pattern, bottom centre, and the Sun on the left-hand side. The Sun's track across the sky is the line below the antenna pattern. In mid-summer, if the Sun reaches 60° or more above the local horizon during transit, this antenna configuration would work well.

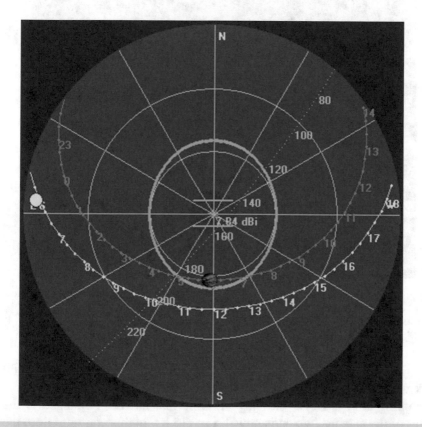

Fig. 12.10 Dual dipole without the 135° phasing cable, at a height of 3 m (10 ft)

Looking at Fig. 12.11, this has the 135° phasing cable in place. Nothing else has been changed; the antenna dipoles are still the horizontal lines in the middle of the image. But look how the shape of the antenna pattern has been changed by fitting the phasing cable. Notice the cross in the centre of the ellipse. Where the maximum gain of the antenna is in this configuration. The maximum gain is at 50° above the local horizon, but look how close it is to Jupiter. This antenna configuration would be suitable for receiving radio emissions from the planet.

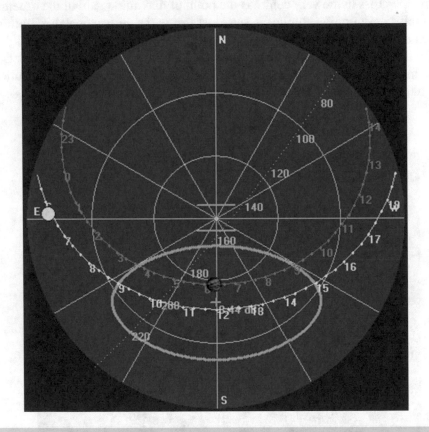

Fig. 12.11 Dual dipole antenna with 135° phasing cable, south-phasing, at the height of 3 m (10 ft)

Notice the track below Jupiter—that of the Sun. The configuration would work well to tune into the radio emissions from the Sun.

Looking at Fig. 12.12, the only variable that has been changed is the height of the antenna, which is now 4.5 m (15 ft). In this configuration, the maximum gain will be at 40° above the local horizon. Look how far Jupiter has moved away from the centre. This change could make all the difference

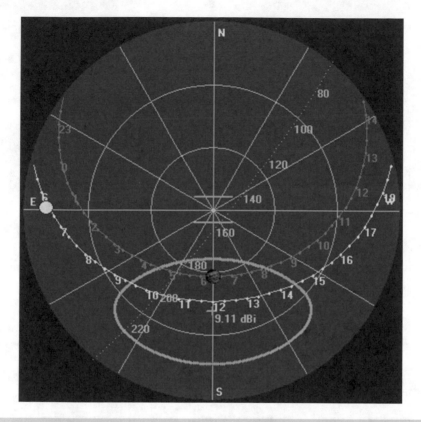

Fig. 12.12 Dual dipole antenna with 135° phasing cable, south-phasing, at the height of 4.5 m (15 ft)

for receiving radio emissions or not. The Sun fares slightly better in this configuration at this time of the year (August in the northern hemisphere).

Figure 12.13 is of the antenna pattern with the antenna set at a height of 6 m (20 ft). At this height, the maximum gain from the antenna will be at 30° altitude above the horizon. If Jupiter is low in the sky, this antenna configuration would get the most gain from the antenna.

As the Sun makes its familiar rise and fall in altitude in the sky over the course of a year, from its highest position in midsummer to its lowest position in midwinter, the height at which the antennas are used will need to be adjusted to keep the object—either the Sun or Jupiter—as close to the centre as possible.

Antenna patterning and gain have already been covered in Chap. 6. Still, let's recap why it is important to try and keep an object as close to the centre of the antenna beam as possible to get maximum gain. Let's take for example a flashlight switched on inside a dark room. Directly in front of the flashlight, there will be a fully lit area projected onto a nearby wall, but all

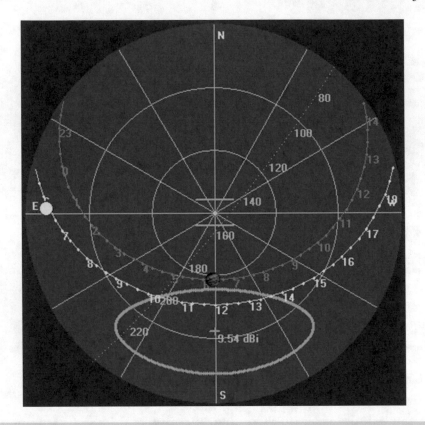

Fig. 12.13 Dual dipole antenna with 135° phasing cable, south-phasing, at the height of 6 m (20 ft)

around this fully lit area will be a semi-lit area. Think of this semi-lit and fully lit area as the antenna beam. If we take Jupiter now, when Jupiter is completely outside the antenna beam, we can't detect it, as it is in darkness. But as the Earth rotates and brings Jupiter across the sky into the antenna beam, it will first enter the semi-lit area. So in theory, the antenna could pick up radio emissions from the planet, but the gain from the antenna will be low. As Jupiter makes its way into the fully lit area of the beam, this is where we get the most gain and this is the best time to listen for radio emissions. As the rotation of the Earth now moves Jupiter out of the fully lit area on the opposite side of the antenna beam, we start to lose the gain of our antenna once again.

Note that the ellipse used to indicate the antenna pattern is a theoretical depiction of the antenna pattern and it is not set in stone. It is therefore best to start observing an hour and a half to 2 h on either side of the transit time marked on the antenna pattern. This is the best chance to pick up the radio emissions.

Radio Jupiter Pro

The software supplied within the kit is fine to get up and running, but its limitations will soon be reached and an upgraded version will be needed. Radio Jupiter Pro is an absolute must-have piece of software to help an observer. It tells the best time to observe and receive radio emissions from Jupiter. This can be downloaded from www.radioskypublishings. It is free to use for the first 30 days then the license needs to be purchased. After purchasing the license, any updates and improvements made to the program are free to download and install, so it is always possible to have the up-to-date version of the program.

Once the software has been loaded, there are a few things that need to be set: latitude and longitude, plus preferred date format and time zone. The program does the rest.

On opening the program, a window pops up showing the relative position of Jupiter and the Sun plus the time and date. At the bottom is a smaller window that informs the user if there is a possibility of any radio noise storms from Jupiter happening at that moment.

Click on the next tab and the *radio storm prediction* window pops up. This window displays a chart showing the predictions for radio storms for the next 24 h.

There are buttons to skip to the next day or so into the future or the past. The predictions are displayed in written form, giving approximate start and finishing times.

The Sun's visibility above the horizon is also shown. Moving the computer's cursor across the screen, the details of the time will change according to the cursor's movements. The positions of the Sun and Jupiter will be shown in degrees of altitude.

The next window is the *Io phase plane*. This charts Jupiter and the central meridian longitudes, giving a probability of receiving a radio storm if Jupiter is in the right part of the sky for the observer.

Look at Fig. 12.14, a screenshot of this particular window. This image may seem strange at first sight, but it is straightforward to understand once a number of key features are pointed out. Jupiter can be seen on the right-hand side just below centre. The lines running from left to right at an angle are Earth times in 24 h. As Jupiter moves along these lines, the probability of picking up a radio storm is seen to change. Looking on the left-hand side of the image, there is a color-coded strip, starting with black at the bottom

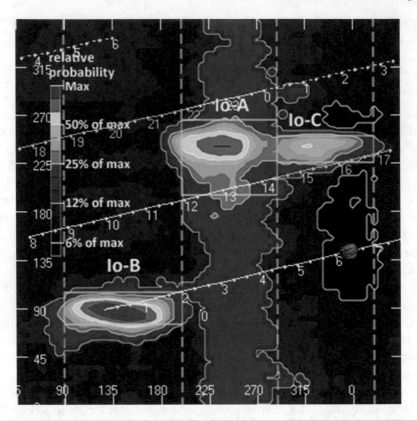

Fig. 12.14 The Io phase plane window in Radio Jupiter Pro

and red at the top. There is also a percentage value marked at different points along the strip to give the user an idea of the probability of picking up a radio storm. Moving the computer cursor across the screen, the probability can be seen to change on the right-hand side of the image.

The Io phase window can be useful, because as Jupiter moves along the timelines, the probability of receiving radio emissions from Jupiter can be seen at any point. The planet crosses the different colored areas on the screen. The brighter the color, the greater the chances of receiving radio emissions. Ideally, Jupiter wants to be in the red area as it passes through the centre of the antenna beam at the observing site.

The next window is the *antenna pattern* window. Recalling the screenshot images from the previous section, this is what this particular window looks like. This part of the program is great for supplying you with the correct time to listen to the Sun and Jupiter, and gives a visual indication of the positions of the Sun and Jupiter within the antenna beam. The program comes with a number of pre-loaded antenna patterns ready for the user to load. This is good, as it shows which configuration is the best for the location, or if coming up with an antenna configuration of your own, you can load it into the program. By moving the computer curser around on the map, the altitude and azimuth coordinates change in the top left-hand side of the screen. This is useful for doing a quick check on the altitude of the Sun and Jupiter.

Personalize the screen by having stars on the background down to magnitude 5, or another magnitude set to match your own location. This can be useful when locating Jupiter in the sky. You may choose to have the galactic plane shown on the map. This particular part of the sky is rather noisy at radio wavelengths and contains some of the "brighter" radio sources as it follows the Milky Way across the sky. This is where Karl Jansky discovered the radio noise coming from the galactic centre in the early 1930s from the powerful radio source Sagittarius A.

The tracks of the Sun and Jupiter are also shown and either one can be switched on or off. A good feature about this window is that it keeps updating itself and shows the progress of the Sun and Jupiter through the antenna beam while carrying out radio observations.

The next window, the *altitude vs. azimuth* window, shows the altitude and azimuth track of the Sun and Jupiter across the sky. Use this window to carry out checks of the altitude of the Sun and Jupiter to gain the best antenna configuration and get the maximum gain from the antenna.

The next window is the *ephemeris* window. This shows a table with the predicted positions of the Sun and Jupiter at given intervals.

The next window is the *plot Jovicentric declination of the Earth* window. This window shows the relative position of the planet Jupiter in the sky over any year the user chooses to set.

The next window is the *yearly visibility schedule*. This window will calculate the best times to observe the planet Jupiter over a given year. For instance, the schedule will indicate when Jupiter will be above the horizon when the Sun has set. The graph is quite easy to understand. Jupiter is shown on the graph as a cyan (blue-green) color, and the Sun is shown in a yellow color. If both are in the sky together, the color changes to a light green, and if neither is above the horizon, the color white will be seen. The months of the year run vertically down the left-hand side, and there are three hourly timelines running from top to bottom. Using this window, you can instantly see the best months of a year to observe Jupiter.

The next window is the *observer's log*. The observer's log shows three different types of radio storm signals: S bursts, L bursts and undetermined. Each one is divided into three different strengths: weak, medium and strong. On hearing a radio emission, for example a medium S burst, just click on the medium S burst button and the program will do the rest. It will log the type of emission, the time and date, and save information about the position of Jupiter in the sky at the time of the storm. This can be saved to a file and studied later or printed out to give a permanent record of the observation session.

The final window in this useful program is the *automated action parameters* window. This allows the user to set specific conditions and tasks for the program to perform, for example to look for a particular type of Jovian storm at a specific date and time and make the user aware of this.

Radio-SkyPipe

Radio-SkyPipe is a very useful program and easy to understand and use. Once loaded onto the computer, it turns the computer into a chart recorder with an on-screen display similar to the old paper chart recorders, but without generating miles of paper tracings. The version that comes with the Radio Jove kit is fine to start with, but sooner rather than later, its limitations will be reached and an upgraded version of the software will be required.

This can be downloaded from www.radioskypublishings. It is free to use for the first 30 days, and then the license needs to be purchased. After purchasing the license, any updates and improvements made to the program are free to download and install, so it is always possible to have the up-to-date version of the program.

This program is capable of recording sound files and playing them back. It is programmable so can be triggered automatically to record sound if a level set by the operator is exceeded. The on-screen instructions are clear and easy to follow. Any files will automatically be saved and are easily retrieved later.

Once Radio-SkyPipe is running, a small control panel will be shown in the top left of the screen. By clicking one of these buttons, changes to the scales of the display can be made. There is also a record button that will immediately start recording a sound file if seeing or hearing something of interest. Notations can be added to the graph so it can be found more easily later on, or even save the screen display straight to a file for retrieving later.

The Solar Cycle

The solar cycle is thought to happen every 9–13 years, with the average being 11 years. This is when the magnetic poles of the Sun reverse. Midway during this change is the solar maximum.

The planet Jupiter has radio emissions that can be predicted with a certain level of accuracy, but the Sun is a law unto itself. Take for instance the lead-up to the 2012 solar maximum, which was very slow in coming. Observers of the Sun for many weeks during the summer of 2011 saw very little in the way of activity. The actual solar maximum didn't happen until early 2014, and this was reported as being one of the weakest on record. Solar astronomers have gone as far as saying the next solar maximum, due around 2024, may be even weaker than the 2014 maximum. Only time will tell.

The Sun has done strange things in the past, for example the Maunder minimum around the year 1715, where solar activity was low for about 70 years. It has been suggested that the total energy output from the Sun also dropped during these years. This reduced output from the Sun has led experts to suggest that this caused the river Thames in London to freeze every winter to such an extent that an open air market and fair could be held on the frozen river.

The theory of the Sun randomly reducing its overall output in the past years has also been linked to large rivers reducing their flow in the years when the solar activity has been low. In other parts of the world, rivers have exhibited the opposite effect, and when solar output returns to normal, so do the river levels. Why this happens is still unknown.

Types of Radio Emissions

Solar activity can play an important role in the radio emissions that the Sun generates.

Radio emissions from the Sun come randomly, although when sunspots can be seen, this is a good indication that there will be some sort of radio emissions. Greater sunspot numbers usually mean more radio emissions, as do large single sunspots, but the Sun will produce the odd surprise now and again. Radio emissions from the Sun can come in a number of different forms and over several different frequencies. Depending on which source of information is used, the description of each type of radio emission can vary slightly.

The following list has been compiled from many different sources, and common ground has been found from each different description.

1. **Quiet Sun:** Taken to mean a low level of solar activity, with little or no sunspots visible, and with X-ray emissions that are below a C-class flare. Other radio emissions are continuous across all wavelengths and originate from thermal radiation. This is an example of the Sun at solar minimum.
2. **Type 1:** Noise storms composed of many short bursts, from a tenth of a second to 15 s. They are of variable intensity and may last for many hours or days.
3. **Type 2** (Slow-drifting bursts): These bursts are caused when a shockwave from a large flare travels up through the Sun's atmosphere. The shock-

wave is caused by material being thrown off by the flare. They are of narrowband emission and slowly sweep from high to low frequencies over several minutes.

4. **Type 3** (Fast-drifting bursts): These bursts are narrowband and drift rapidly from high to low frequencies over several seconds. They are associated with active regions of the Sun's surface, for example a solar flare or large sunspot.

5. **Type 4**: Continuum emissions can last from many hours and are associated with major flare events, picked up after the flare has erupted and reaching its maximum intensity.

6. **U-burst**: Sometimes called *Castelli-U*, they can last for several seconds and the wavelength changes rapidly, decreasing and increasing again. They are often associated with flares and have a similar origin as Type 3 active regions on the Sun's surface.

Figure 12.15 is an example of solar activity. This was received by using the Radio Jove receiver with its dual dipole antenna using Radio-SkyPipe software.

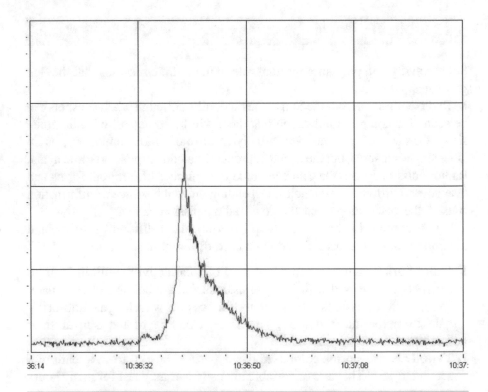

Fig. 12.15 Single Type 3 solar burst

This is a Type 3 burst—note the burst's classic shape of a shark's fin. Note also the rapid rise (the almost vertical line on the left-hand side) and the slow return to background again (the shallow-angled right-hand side).

Here is link to audio examples of solar radio emissions like the one above: www.radiojove.gsfc.nasa.gov/observing/sample_data.htm.

Jupiter's Magnetic Field and Magnetosphere

If it were possible to see Jupiter's magnetosphere from the Earth, it would look the same size as the full Moon. It is the largest entity in the Solar System, even bigger than the Sun itself. The intensity of the magnetic field measured from the cloud tops of Jupiter is in the region of 14 times that of the Earth. This may not sound like a great difference but if the size of Jupiter is taken into consideration, it makes a huge difference. The secret to this phenomenal magnetic field and magnetosphere is contained within the internal structure and composition of Jupiter itself.

Jupiter is sometimes referred to as a failed star. Theories have suggested that if Jupiter was approximately ten times larger, the gravitational forces developed by its mass would be great enough at the core to start nuclear fusion, not at the rate that happens at the centre of the Sun, but as a brown dwarf. There is some conjecture as to whether this would be true or not. Jupiter radiates twice as much energy as it receives from the Sun. There are many theories why, the leading one being that Jupiter's solid core is slowly shrinking and producing internal heating.

Starting from the outside and working towards the core, we will see how Jupiter generates its powerful and extensive magnetosphere. The planet's atmosphere is mainly a mixture of hydrogen and helium, with a small percentage of other elements. The atmosphere extends many thousands of miles down to its solid core. Traveling through the atmosphere, the atmospheric pressure increases, and we reach a transition point where hydrogen ceases to be a gas and starts to change into a liquid. Traveling further down, all the hydrogen gas has gone and there is just liquid hydrogen. Traveling still further, we reach another transition zone where the atmospheric pressure is now so great that the hydrogen liquid is starting to turn into a more exotic liquid—that of metallic hydrogen. Travel even further, we reach the point where all the liquid hydrogen has turned into liquid metallic hydrogen. The last stop on the journey is what is thought to be a rocky core.

Jupiter owes its magnetic field to the large amount of liquid metallic hydrogen in its outer core. The gas planet has the same convection currents in its outer core as the Earth. If we then take into account Jupiter's size and faster rotation period of <10 h, this makes a very powerful dynamo that can produce a magnetic field 20,000 times more powerful than the Earth. This results in Jupiter's vast magnetic field and magnetosphere.

Radio Emissions and the Io Effect

Radio emissions from the planet Jupiter are divided into two types: L bursts (L = Long) and S bursts (S = Short). *L bursts* have been described as gentle waves washing up on a sandy beach, or a breeze blowing through a leafy tree. *S bursts* have been described as pebbles on a beach being moved around by the action of the sea, while others think they sound like the sizzling of bacon from a hot frying pan. Both these sounds can be heard separately or together, but they are always heard with white noise in the background. Radio emissions can also be produced within the magnetosphere by particles from the solar wind getting trapped around magnetic field lines, spiraling them down and releasing energy in the form of radio waves.

Jupiter's innermost moon, Io, is the most volcanic world in the Solar System. It orbits very close to the planet and is well within the Jupiter's magnetic field. It has been suggested that the action of Io orbiting at such a close distance causes an electrical current to be produced. This current has been estimated to be somewhere in the order of a billion amps. This means that if it were possible to stand on the surface of Io, the electric current would make for a very good show of aurorae in Io's wispy sulphurous atmosphere, but the show couldn't be enjoyed for long before receiving a lethal dose of radiation from Jupiter!

Io has a part to play in these radio emissions. The phenomenon is called the *Io effect*. As Io orbits Jupiter, all the electrical energy produced has the effect of concentrating any radio emissions into a large cone shape. The cone travels ahead of Io in its orbit, focusing any radio emissions into a sweeping beam, similar to that of a lighthouse. If Earth is in the path of the beam, any radio emissions will be of greater intensity than if Io was at another point in its orbit.

To hear audio samples of Jupiter, visit the Radio Jove website here: www. radiojove.gsfc.nasa.gov/observing/sample_data.htm.

The Best Time to Listen to the Sun and Jupiter

The Sun

The Sun can be observed at any time as long as it is within the antenna beam. The best time to try and receive signals from the Sun is when there is a lot of solar activity, such as sunspots and solar flares. Any time around solar maximum is a good time, though there is no need to wait for this period to start observing.

Once the equipment is set up, tune the Radio Jove receiver to a clear frequency. During the day, the shortwave frequency range may be busy with radio stations, and it is not uncommon to have to re-tune the receiver a few times to find a clear frequency to receive the Sun. Re-tuning won't make a lot of difference for picking up the Sun, as radio emissions are received over a number of frequencies simultaneously.

Here is a link to a great website that covers just about everything to do with the Sun and space weather: https://www.swpc.noaa.gov/. Once there, just follow the links to other useful sites.

Jupiter

Picking up radio emissions from Jupiter needs a little more planning, as we need to take into account the Earth's ionosphere. Looking back to Chap. 4 about the Earth's ionosphere and referring to a SuperSID Fig. 4.2 showing the sunrise and sunset, this image shows how quickly the ionosphere changes during a 24-h period. The sunrise side of the graph shows an almost vertical line. At this time, the ionosphere is ionized almost immediately as the Sun rises and the ultraviolet light hits it. Now look at the sunset side of the graph. Here, the line is at more of an angle, showing that the ionization took 2 or 3 h to fade. If solar activity is high, like at solar maximum, the ionization of the ionosphere may take most of the night to fade away. If there is auroral activity, the ionization will be high all night and never fade. This acts like a barrier to some radio frequencies, and the 20 MHz used here is one of those frequencies that are blocked. Fortunately, radio emissions from the Sun are strong and can "punch" through this barrier, but radio emissions from Jupiter are weaker and cannot do the same.

This means as a rule, radio emissions from Jupiter can only be picked up at night. Although some observers have managed to pick up Jupiter during the day, these observations are rare and need conditions that are just right. The Sun must be low down in the sky with very low solar activity, and the planet Jupiter must be high in the sky.

Here is how to tell whether it will be a good night to pick up radio emissions from Jupiter. Once the equipment has been set up, if a large number of radio stations can be heard and it is difficult to find a clear frequency, this indicates the ionosphere is still heavily ionized. In this situation, leave it for an hour or so, then try again. If things have quieted down, look for an empty frequency.

Jupiter's radio emissions are weaker than those of the Sun and therefore require a radio-quiet environment in which to receive them. It may not be possible to pick up Jupiter if the equipment is used in a radio-noisy environment. This is the case for some observers, but their equipment still manages to pick up the stronger radio emissions from the Sun. A good test to find out if the observing site is radio-quiet is to try and pick up the galactic centre as Karl Jansky did in 1932. If this can be received, picking up Jupiter shouldn't be a problem.

This can be seen by visiting the Radio Jove web site. Here is the link: www.radiojove.gsfc.nasa.gov/observing/sample_data.htm.

Useful Books and Podcasts

Listening to Jupiter: A Guide for Amateur Radio Astronomers, Second Edition. By Richard S. Flagg. 2000.

This book is highly recommended. It is written in a friendly and easy-to-read manner, and there are several funny anecdotes, such as the one about how flames shot out of a project the author had recently finished building, and one about the night he came across an alligator. It is included as a PDF copy through a Dropbox link when you purchase the Radio Jove kit.

There is a podcast about the Radio Jove project and receiver available here: https://cosmoquest.org/x/365daysofastronomy/?s=Radio+Jove+project.

Radio Jove Links

Below are links to Radio Jove online material, kindly supplied by Mr. Chuck Higgins, one of the Radio Jove mentors.

- The Radio Jove Primer Education CD: https://radiojove.gsfc.nasa.gov/education/educationalcd/
- Receiver and antenna construction manuals and soldering videos: https://radiojove.gsfc.nasa.gov/telescope/equipment_manuals.htm
- Radio Jove brochure: https://radiojove.gsfc.nasa.gov/library/brochure.pdf

Education Materials (Somewhat Dated)

- Education Materials: https://radiojove.gsfc.nasa.gov/education/materials.htm
- Lesson Plans: https://radiojove.gsfc.nasa.gov/education/lesson_plans/lesson_toc.ht
- Classroom Exercises: https://radiojove.gsfc.nasa.gov/education/exercises.htm

Software

- Radio Jove Software: https://radiojove.gsfc.nasa.gov/software/
- Direct Link to Radio-Skypipe: http://www.radiosky.com/skypipeishere.html
- Direct Link to Radio-Jupiter Pro: http://radiosky.com/rjpro3ishere.html
- Direct Link to Radio Spectrograph: http://jupiter.wcc.hawaii.edu/spectrograph_software.htm

Part III

Radio Astronomy with a Software-Defined Radio (SDR)

Chapter 13

The Software-Defined Radio (SDR)

What Is a Software-Defined Radio?

Regardless of its modulation, a radio signal can be looked upon as an alternating current (AC), granted a very small one. This alternating current is therefore an analogue signal.

A Software-Defined Radio (SDR) in it simplest form is an analogue-to-digital converter capable of working at radio frequencies. This digital signal is then fed into a computer and the software processes the signal into audio and a visual display on the computer screen.

The idea of the SDR isn't new; it goes back to the mid 1980s. The technology for the SDR came into being in the early 1990s. The first commercial SDR was available in the late 1990s and into the early 2000s.

How the SDR Works

The best way to understand how the SDR works is to compare it with an analogue radio. For clarity, simple block diagrams and an explanation of what happens at each stage will be adequate.

In the case of the analogue radio (Fig. 13.1):

S. Arnold, *Radio and Radar Astronomy Projects for Beginners*, The Patrick
Moore Practical Astronomy Series, https://doi.org/10.1007/978-3-030-54906-0_13

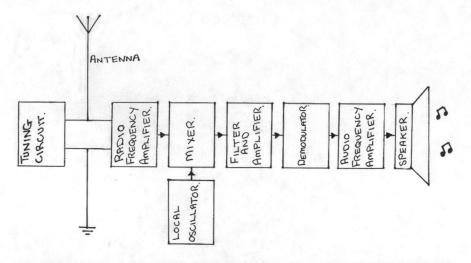

Fig. 13.1 Block diagram showing how an analogue radio works

1. The antenna picks up all radio frequencies.
2. The wanted frequency is selected through a tuning circuit.
3. The signal is then passed into a radio frequency amplifier to boost the signal.
4. The signal is then fed into a mixer that mixers the radio signal with a signal from a local oscillator. The purpose of the local oscillator is to change the frequency of the radio signal, which improves the performance of the radio. The oscillator is usually a crystal, as these produce very stable frequencies and are not affected too much by temperature changes.
5. The output from the mixer is then fed into a filter to remove the radio frequency from the original signal. We are then left with the modulated audio signal.
6. The modulated audio is fed into a demodulation unit that removes the modulation from the audio signal. We are now left with the audio we want, but at a low level.
7. The output from the demodulation unit is passed into an audio amplifier to boost the power enough to drive a speaker.

As can be seen from this simple explanation, the original radio signal passes through a number of steps to remove the audio from it. At each stage, the analogue radio uses built-in circuitry with electrical components to filter, demodulate and amplify the signal. At all stages, the signal remains an analogue one.

If we now compare the workings of the analogue radio with the workings of a SDR, we shall see the difference in how they work.

In the case of the Software-Defined Radio (Fig. 13.2):

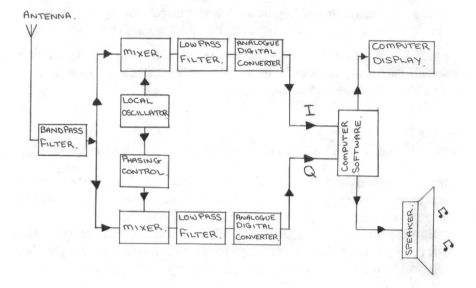

Fig. 13.2 Block diagram of a Software-Defined Radio

1. We still have an antenna to pick up the radio signal.
2. The radio signal is now passed through a bandpass filter. This is the equivalent to the tuning circuit on an analogue radio. A bandpass filter works by only letting frequencies through that have been set between two values and rejecting everything else.
3. The input is now split into two the I input and Q input.
4. We now have two mixer stages instead of the single one on an analogue radio. One is for the I input and the other for the Q input.
5. We still have a local oscillator, but instead of a crystal, the SDR uses an oscillator in conjunction with the computers processor. The output from the local oscillator now feeds into both the I and Q inputs. On the Q side we have a phasing shift control. This is used to change the phasing of the input signal in a similar way to that described in Chap. 9 Quadrature Modulation.
6. Both the I and Q feeds are then passed through a low pass filter. These filters work by only letting through frequencies lower than that which the filter has been set.
7. Then both the I and Q feeds are passed into an analogue-to-digital converter. At this stage, the signal is purely a digital one.

8. The output from the SDR is fed into the computer via a USB connection.
9. The SDR software takes the digital signal and using mathematics and algorithms demodulates the signal in the way chosen by the user. The software uses the computer to produce a visual display of the signal and the computer sound card to amplify and drive the speakers to produce audio.

In the SDR, the software really does define the radio, as the software does all the work. Without it, there is no radio. SDR's are now used on everything from satellites to smartphones.

Types of SDR's

There are several different types of SDR's on the market. They start in price from around $10 (£6) up to a few hundred dollars. The four that will be discussed here range in price from around $10 (£6) to $55 (£33).

The first one is the mini DVB RTL-SDR. It costs around $10 (£6). This SDR dongle is cube-shaped and approximately 13 mm (0.5 in.) square. This SDR dongle is a fake. It was advertised as having a R820T2 tuning chip but in reality it has the inferior FC0013 tuning chip.

The frequency range of the FC0013 tuning chip is up for debate but is thought to be about 48.25 MHz to 863.25 MHz, unlike the R820T2 tuning chip that covers frequencies from 25 MHz to 1.7 GHz. It comes with a built-in USB connection and a MXC (Micro CoaXial) input connection, as well as a 100-mm (4-in.) antenna that claims to be able to receive television signals (in Europe). In reality, this antenna is all but useless and struggles to pick up the local FM radio station. Fitting it to a 1.4-m (54-in.) vertical CB antenna improves the performance of the SDR dongle, but the tuning accuracy of the FC0013 chip is pretty awful. This particular one shows a difference of 25 KHz when tuning into the local FM radio station. This could prove problematic when trying to calibrate the SDR software. Avoid this type of SDR dongle as it is not worth the money (Fig. 13.3).

The second one is the PAL/Belling-Lee RTL-SDR. It cost around $10 (£6). This SDR dongle comes with a built-in USB connection and a female coax connection at the input. Its frequency range is said to be 25 MHz to 1.7 GHz, but in reality the coax connection is not suited to the high frequency. Its performance is poor and the signal starts to get noisy around the 150 MHz range. This SDR receiver is not recommended (Fig. 13.4).

The third is the Andoer MXC (Micro CoaXial) connection dongle SDR. It costs around $12 (£8). Several different companies manufacture

Fig. 13.3 Mini DVB RTL-SDR

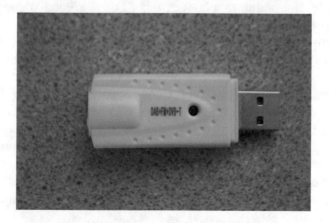

Fig. 13.4 Belling-Lee RTL-SDR

this type of SDR receiver but beware of fake tuning chips. This SDR dongle was first made to receive television signals in Europe before people quickly realized it works well as a SDR. This SDR comes with a built-in USB connection and a MXC input connection. It comes with a 100-mm (4-in.) antenna that is pretty much useless except for the most powerful local FM radio stations. This particular one comes with software to watch digital television and a remote control.

Its frequency range is 25 MHz to 1.7 GHz. Its performance is certainly better than the previous ones, but it does suffer from interference due to the

Fig. 13.5 Andoer mini portable dongle

plastic body. If working to a budget, this SDR is a good starting point or spare (Fig. 13.5).

The forth one is the Auzeuner V3.0 SDR. It cost around $28 (£19). This SDR comes with a short USB cable to connect the receiver to the computer, which is a good idea as it takes the weight off the computer connection and isn't so easily knocked like the previous three. The input is a SMA (Sub Miniature version A). Its frequency range is 25 MHz to 1.7 GHz. The V3.0 version is marketed as having more accurate and stable electronics. Its performance is very good and it seems to be very stable when running. It is totally enclosed in an aluminum case that shields the internal electronics. If already running SDR software, an updated driver download maybe required. This SDR receiver performs very well and is recommended (Fig. 13.6).

Fig. 13.6 The Auzeuner RTL-SDR V3.0

The fifth is the Excellway Ham radio receiver SDR. It cost around $50 (£33). This SDR comes with a short USB cable to connect the receiver to the computer and a 300-mm (12-in.) antenna that performs slightly better than the 100-mm (4-in.) one mentioned early. It's entirely enclosed in an aluminum case that shields the internal electronics.

The best advantage with this SDR is its frequency range. Looking at Fig. 13.7, it can be seen that there are two inputs. One input covers the frequency range from 100 KHz to 30 MHz and the second covers the frequency range from 25 MHz to 1.7 GHz. This frequency range makes its suitable for receiving radio emissions from Jupiter and the Sun. One serious drawback, however, is the USB cable supplied with it. In a word, it's rubbish, which is a shame as this could put people off. The cable has very poor shielding qualities and picks up interference like a second antenna. Once the shielding is sorted out, the SDR performance is greatly increased. This is why despite this drawback, this SDR receiver is highly recommended.

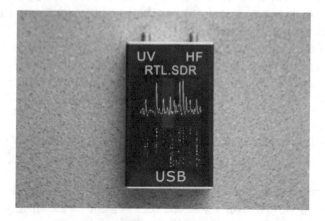

Fig. 13.7 Excellway Ham radio receiver

There are higher priced SDR receivers, for example the "Funcube" dongle, that have a greater frequency range, may have better electronics and in some cases have better software. But the SDR receivers discussed here offer excellent value for money and if you are new to the subject, these are a good starting point.

Caution when Purchasing SDR's

Ideally, go to a radio supplier and ask somebody for advice, as they may even help set up the SDR software. Nevertheless, chances are that the SDR will be purchased online, so here are a few things to watch out for.

1. Some SDR receivers (like the one shown in Fig. 13.7) can be purchased in kit form to build it yourself. These kits are a few dollars cheaper. Speaking from experience, pay the extra and get one ready-built. It is an absolute nightmare to build some of the components used in SDR receivers, as they are really small—some even being half the size of a grain of rice.
2. Use a reputable seller. If unsure, ask questions. Any reputable seller will always reply to questions. Beware of oversea cloned or fake SDR's.
3. Beware of the "stock photo" scam. This is used a lot when buying second-hand books. An image of a pristine book may be shown, but when it arrives, it may look like a dog has been chewing it. When ordering online, download an image of what's been ordered and make sure it's the same item when it's delivered. Make sure it is not a supplier showing an image of the latest SDR receiver and sending out their old stock.
4. Look at customer feedback, especially when buying from overseas. Not all negative feedback is bad. It could be someone complaining it was delivered a day or so late. Look for things like "item never arrived" and "seller never replies to emails."
5. Some of the more expensive SDR receivers will go as low as 1 MHz and work perfectly well. Some of the cheaper SDR receivers have started to be advertised to go as low as 1 MHz, but when tried, they cannot go below 25 MHz. Be careful of false frequency claims.

How to Tell a Genuine from a Fake

Buying things online is a good way to save money. There are lots of good, honest sellers who deliver excellent service, but unfortunately, there are some less honest sellers who advertise genuine items.

There are some fake SDR dongles on the market. These are usually but not always the plastic ones. A genuine dongle uses an R820T2 tuning chip, while a fake dongle uses an inferior FC0013 tuning chip. These fake SDR dongles give poor performance, poor tuning accuracy, their frequency ranges are smaller and they have gaps within their frequency ranges. Even if an SDR dongle is advertised as having an R820T2 tuning chip, on closer examination, some do not.

If The SDR dongle is delivering poor performance, if there is difficulty in tuning the dongle using the software or if you suspect the SDR dongle is fake, there are four ways to check.

The first and probably the quickest and easiest way is to use HDSDR software, which works on all the above SDR receivers. Plug in the SDR receiver and open the software. Then click on the "ExtIO" button. A window will open showing the SDR tuning chip on the right-hand side. If more than one SDR receiver is being checked, close the software, remove the previous SDR receiver, then plug in the one to be tested before opening the software again.

The second way is to use SDRsharp software, which also works on all the above SDR receivers. Plug in the SDR receiver and start the software running. Click on the control icon top left, and a window will open. Look at the very top at the "Device." At the top right will be found the number of the tuning chip contained within the SDR receiver. As above, close the window before testing another SDR receiver.

The third way is to use another piece of software. Here is a link to a website with instructions on how to use the software and a link to download the software itself. Before downloading the software, take a moment to read how to use it: https://inst.eecsberkeley.edu/~ee123/fa12/rtlsdr. html.

The software is downloaded as a ZIP file, so look for a ZIP file with the name "rslsdr_win." Open this file and install the software. Plug in the SDR dongle and open the software. A small window will appear on the screen. Let the software run through its tests, and the details of the dongle will be shown in the window. Look about 2/3 down the list at the "tuner found." If it shows R820T2, the SDR is genuine. If it shows FC0013, the dongle is a fake.

The fourth should only be done if the above methods fail. This is to check the internal circuitry of the SDR dongle, but caution should be taken as this could void any warranty that comes with the SDR dongle. If the dongle is a plastic one, this can be done by using a small flat-bladed screwdriver. Looking around the edge of the dongle, a seam will be seen. Push the tip of the screwdriver in this seam and slowly work around the edge of the dongle. Carry on working around until the plastic case opens. If the SDR dongle is in a metal case, there are four cross-head screws at either end of the case. Undo these screws and the circuit board can be slid out. Once the circuit board has been removed, orientate the dongle until it matches the one shown in Fig. 13.8.

Look at the chip shown in the image below—a magnifying glass will be needed. The number on the chip should be marked R820T2. If not, the dongle is a fake.

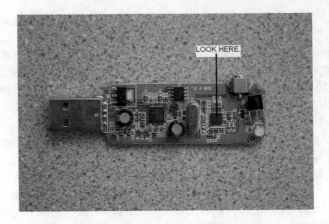

Fig. 13.8 Image showing how to spot a fake SDR dongle

Refit the circuit board in the reverse procedure as it was removed. The plastic dongle will just "click" back together.

Chapter 14

SDR Software

Computer Operating System

There are several different versions of SDR software, available either as license-paid or freeware. The best thing to do is try the freeware, and if a more advanced version is required, look at some of the paid versions. The two that will be discussed here are both freeware.

Before discussing the software, the first thing we need to find out is what your computer system is, as nearly all SDR software require a 64-bit system to run properly. On a Windows operating system, right click on the windows button on the bottom left. In the search bar, type in the word "System" and click on the item listed beneath "Control Panel." In Fig. 14.1, you will see that the computer runs Windows 10 with a 32-bit operating system.

If the system is 64-bit, great: this will run 64-bit and 32-bit software, and there are several versions of SDR software to use. If the system is 32-bit, this narrows down the choice of software considerably, although there is one really good 32-bit SDR software available.

S. Arnold, *Radio and Radar Astronomy Projects for Beginners*, The Patrick
Moore Practical Astronomy Series, https://doi.org/10.1007/978-3-030-54906-0_14

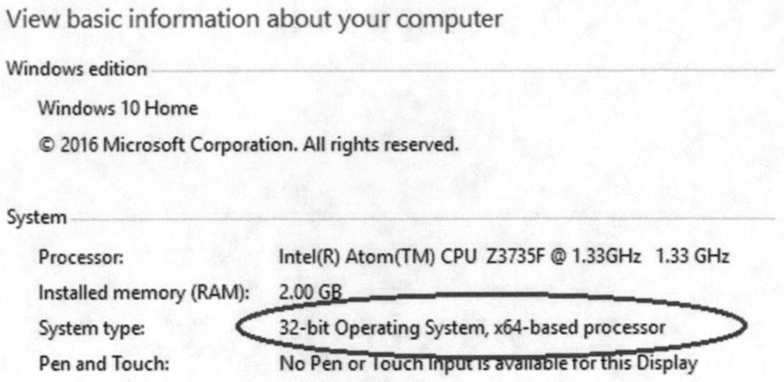

View basic information about your computer

Windows edition

Windows 10 Home

© 2016 Microsoft Corporation. All rights reserved.

System

Processor: Intel(R) Atom(TM) CPU Z3735F @ 1.33GHz   1.33 GHz

Installed memory (RAM):   2.00 GB

System type: 32-bit Operating System, x64-based processor

Pen and Touch: No Pen or Touch Input is available for this Display

Fig. 14.1 Image showing 32-bit operating system

SDRsharp Software

SDRsharp is 64-bit software and very popular with SDR users. It can be downloaded from a number of sites, including:

* https://www.rtl-sdr.com
* https://airspy.com

The SDRsharp manual can be downloaded from the same sites. The software is downloaded as ZIP file. Extract the files in the ZIP file to your computer. Then open the file and look for the "install" file. Double click on this file and the computer will start to install the software. The usual warning messages will pop up—just click on "yes" or "allow." The computer may ask to have *Microsoft.NET4.6 redistributable* installed to use a SDR. If so, this needs to be downloaded.

Plug in the SDR and wait for the computer to install any software. Go back to the file and click on the "zadig.exe" file. A window will pop up; click on the options button. Click on all devices. Look for "Bulk-In, Interface (Interface 0)." Then click on the "replace driver" button. Open the SDRsharp software and the window in Fig. 14.2 is shown.

Click on the source button at the top and a list will be shown. Click on "RTL-SDR (USB)." Then, looking top left, click on the icon third from the left (it looks like a gear wheel). A list will be shown. Scroll down and look for "Genetic RTL-2832U OEM." Then click on Sample Rate scroll down and select "2.048MSPS." Next, click on "Sampling mode" scroll down and click on "Quadrature Sampling."

That covers the installation of the software. If the software doesn't run, rerun the above steps. If this doesn't work, here is a link to a video showing

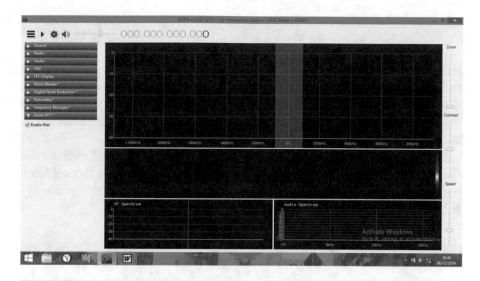

Fig. 14.2 Image showing SDRsharp software window

how to install the SDRsharp software: http://www.youtube.com/watch?v=4Vo3CSU0NFY.

It's time to try the software. First, find the frequency of the local radio station and its modulation, whether AM or FM. Commercial radio stations broadcast on *Wide Frequency Modulation (WFM)*. This can be set by clicking on "radio" button and click on WFM.

The Fig. 14.3 shows SDRsharp software tuned into BBC radio 2.

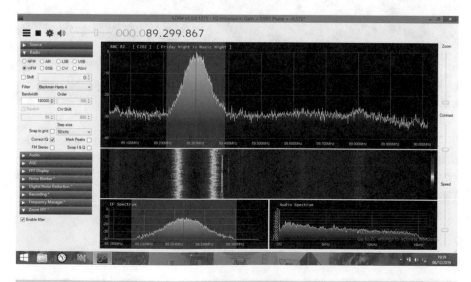

Fig. 14.3 Image showing SDRsharp tuned into BBC radio 2

HDSDR Software

HDSDR is a cracking piece of software and ideal for working on a 32-bit operating system. Download it from the homepage http://www.hdsdr.de. The instruction manual can be downloaded from this site too. Depending on the operating system of the computer, HDSDR software is far easier to get up and running than SDRsharp.

First thing is to download the HDSDR file. Open the file and click on the "install" file, then let the computer install the software. Once the software has been installed, double click on the desktop shortcut and the display should look like Fig. 14.4.

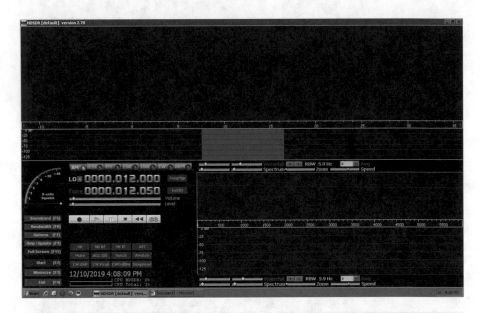

Fig. 14.4 Image showing HDSDR software window

To get HDSDR working, we need to manually copy and paste in the driver for the SDR. Start by downloading SDRsharp as previously discussed. Click on the ZIP file and click on "extract all." Open the SDRsharp file and look for the driver as in Fig. 14.5.

Fig. 14.5 Image showing the SDR driver

Once found, right click and copy. Right click on the HDSDR desktop shortcut and left click on "open file location," then paste the driver in this file. Then, delete the SDRsharp software from the computer if it is a 32-bit system, as it will not run it. Restart the computer.

Plug in the SDR. Double click on the HDSDR desktop shortcut to open the software. Look for the button ExtIO just above the volume control. Right click on the ExtIO button, and the driver just pasted will be there. Using the frequency setting (left click to change up and right click to change down), set the frequency to the local radio station. Set the modulation to either AM or FM. If it's a FM station, start the software running. If it sounds garbled, this is because using HDSDR, the bandwidth has to be set manually. This is done by clicking on the "Bandwidth" button and selecting the 192000, or using the slider control marked FM-BW.

The radio station should sound alright now. Figure 14.6 shows the HDSDR software tuned into BBC radio 2.

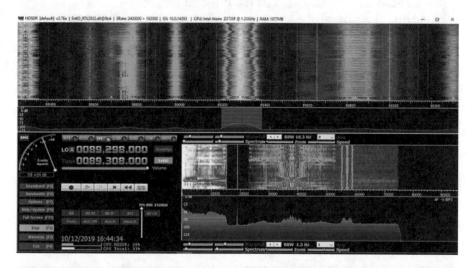

Fig. 14.6 Image showing HDSDR tuned into BBC radio 2

Using SDR Software

Whatever software—SDRsharp or HDSDR—is chosen, both have roughly the same controls. Some controls are easier to use on one software or the other, and it boils down to personal preference. Ideally, if the computer has a 64-bit operating system, download both. The best thing to do is experiment with the software.

Try tuning into different radio stations. This is quite easy; just click on the frequency on the waterfall display to select that frequency. Work through the frequency range of the SDR to see what signals can be picked up. Try using the different modulation settings on these signals, taking note of how the different modulations change the display. One thing to try is picking up the signal from a remote locking control for a car. Operate the remote control and scan through the radio frequencies to look for the signal. Once found, if AM modulation is selected, an audible sound will be heard. Take note of this frequency, and if the remote develops a problem such as a flat battery, it can soon be checked.

Also try at different times of the day. Radio hams sometimes come on the air in the evening, and it can be quite interesting listening to them discussing radios and other equipment.

Using the sliders and zoom controls, see how they change the display.

At this stage, all the filters and other controls can be left at their default setting. These will be discussed as the projects progress.

Soon, the limit of what can be achieved using the supplied antenna will be reached.

If it is possible to connect the SDR to an external antenna, more stations can be picked up. If not, a good starting antenna is a 1.4-m (54-in.) CB antenna. These can be picked up relatively cheaply at garage sales or online auction sites. This will greatly increase the amount of stations that can be picked up. It will also allow ham radio signals to be received, as well as the ISS.

Before long, it will be possible to look at the "waterfall" display and know what modulation to use, for example WFM or NFM, and also whether music or speech is being broadcast. Once familiar with the basic controls of the software, for example changing the frequency and modulation controls, it's now time to calibrate the software.

Computer Processor

When listening to local FM radio stations, the signal is strong and clear, so calibration isn't so important, but if we are trying to find a weak signal, we need to know exactly where to look and at what frequency. As mentioned earlier, the SDR software is the heart of the SDR receiving system and uses mathematics to perform its functions such as accurate frequency display and what modulation to use.

The computer that runs the SDR software has an internal processor that runs the operating system of the computer. This processor acts like a clock controlling the internal operations. The faster the processor, the faster the computer. The processor speed is usually measured in GHz and is the number of functions that can be carried out by the computer every second.

Let's take a computer that has a processing speed of 1.75 GHz. In reality this processor may run at a slightly slower speed of 1.74 GHz or slightly faster at 1.76 GHz. This is perfectly normal, as two processors will never have the exact same operating speed. This isn't noticeable when using a computer for every day tasks, but with SDR software, this difference becomes an important factor. A SDR has an internal oscillator. Two SDR's of the same model can have oscillators that run slightly faster or slower than each other, and this needs to be accounted for. To operate correctly, the SDR software needs to know about any discrepancies in the operating frequencies of the computer's processor and the SDR's internal oscillator.

Calibrating SDR Software

This is an easy operation to carry out, although the two pieces of software, SDRsharp and HDSDR, have slightly different ways of doing it.

First, a suitable signal transmitter needs to be found. A local FM or AM radio signal can't be used as the bandwidth is too wide. It needs to be a very narrowband signal—ideally a continuous wave (CW) signal. This can be anything, like a timing signal, WWV signal or a high frequency beacon. There are high frequency beacons all around the world. Some of these beacons are operated and maintained by amateur radio hams.

Here is a link to a list of world-wide high frequency radio beacons. It can be downloaded as a PDF file: https://iaruhfbeacons.files.wordpress. com/2019/07/worldwide-list-of-hf-beacons-july-2019.pdf.

Calibrating SDRsharp Software

Start the software and tune into the frequency of the chosen signal. Using the zoom slider on the top right-hand side, zoom into the signal. At this point, the display will become very poor in resolution. On the left-hand side, change the resolution setting until the signal becomes clear. Now, click on the centre of the peak. This will show how much the calibration needs to be moved. For example, if the calibration frequency is 150.000.000 MHz, look at the top left: it may be slightly higher at 150.000.050 MHz or slightly lower at 149.999.950 MHz.

Next, retune the frequency to 150.000.000 MHz. Click on the setting button top left. When the window opens, look at the bottom for the "frequency correction (ppm)" control. Click on the up/down control and watch how the signal changes. If the signal moves away when clicking up, then click down. Keep going until the red indicator line passes straight down the centre of the signal's peak. Close the control window. The software is now calibrated.

Here is a link to a video showing the above steps on how to calibrate the SDRsharp software: https://www.youtube.com/watch?v=IaKEYEyrRgk.

Calibrating HDSDR Software

Start the HDSDR software and tune into the chosen calibration frequency. Look for the signal on the spectrum display. Use the zoom slider to zoom into the signal and get a clearer view. To change the screen resolution, use the RBW control, clicking right to increase it and left to reduce it. Click on the centre of the signal so the red line passes straight down the centre of the peak.

Now click on "Options" button on the left-hand side of the screen. A window will pop up. Look for "calibration settings" and click on this. A second window will pop up. Look for the "Correct Tune Frequency (Hz)" box. In this box, type in the correct frequency on the calibration frequency (note this is in Hz), so using the above example of 150 MHz, type in 150,000,000. Then press the "OK" button bottom left. The software is now calibrated.

Here is a link to a video showing the above steps how to calibrate the HDSDR software: https://www.youtube.com/watch?v=PmkATQn3dw0.

A few things to remember when calibrating SDR software:

1. Make a note of the frequency used to make the calibration, as this will be needed again.
2. The calibration needs to be checked periodically.
3. Recalibration will be needed if the SDR receiver or the computer is changed.
4. A good tip to get a really accurate calibration: It will be noticed that the SDR receiver gets warm when operating. This is normal. Before making any changes to the calibration, run the software at the calibration frequency for 10 min before calibrating. This will allow the internal electronics of both the computer and SDR receiver to warm up and stablize before any changes are made to the calibration.

Recording with SDR Software

Before making any recordings, here are a few words about which format to use. Don't use MP3 format for recording. This is fine for pop music, but for more sensitive recordings, it's a poor substitute to a WAVE or WMF file. MP3 recordings use an algorithm that picks up certain frequencies and leaves others out. This is why MP3 format are disliked be people who listen to classical music, as it misses the subtle changes within the music.

On the HDSDR software recording is quite easy—just press the record button and the software will start doing so. The software will record (as default) as a WAVE file. To replay the recording, press the play button. This will open a window and show the recorded WAVE files.

If recording on SDRsharp, the default is a WAVE file, but by default the setting is 16-bit. At this level, 16-bit will eat through the memory on the computer quite quickly. Eight-bit recordings are more than adequate for most purposes. This will need to be changed each time the software is opened. On SDRsharp, we have the choice of recording the audio or baseline. When recording, check the audio box and uncheck the baseline, and set it to 8-bit. To play the recorded audio, click on the "Source" button top left, then click on IQ file (*.wav). A window will open, and the WAVE file can be played back through SDRsharp. An important thing to remember: Don't forget to change back to the SDR driver on the source selector, or the SDR won't work when plugged back in.

Chapter 15

Picking Up the International Space Station

The International Space Station

The international Space Station (ISS) to date is the most expensive single item ever built.

The project is a collaboration involving five space agencies: NASA of the USA, CSA of Canada, Roskosmos of Russia, JAXA of Japan and ESA space agencies from around Europe.

The ISS itself comprises of a number of pressurized modules. It has vast solar arrays to supply it with power. The function of the ISS is to be a space laboratory and observatory. On board, the astronauts carry out scientific experiments in physics, biology, astronomy and meteorology.

The construction of the ISS started in the late 1990's, when the first module was sent into space on board a Russian Proton rocket on 20th November 1998. This first module was to provide propulsion, attitude control, electrical power and communication. Later, more modules were sent into space and bolted to each other. The ISS is also the largest artificial body in orbit. Measuring 109 m (357 ft) in length and weighing nearly 422,727 kg (930,000 pounds), it is around four times the size of the Russian space station Mir and five times the size of the United State's Skylab. It orbits the

S. Arnold, *Radio and Radar Astronomy Projects for Beginners*, The Patrick Moore Practical Astronomy Series, https://doi.org/10.1007/978-3-030-54906-0_15

Earth at an altitude of between 330 km (205 miles) and 435 km (270 miles), and travels at a velocity of around 27,835 kmph (17,300 mph). At this velocity, the ISS completes 15.6 orbits a day and treats those on board to a sunrise and sunset every 92 min.

When seen with the unaided eye at night, the ISS appears as bright as the planet Venus and slowly moves across the sky. Due to the orbit of the ISS, it can be seen from almost any point on the Earth's surface at some time or other, weather permitting.

A Test Run

Before trying to pick up the ISS, a good idea is to have a practice run trying to pick up some local amateur radio hams broadcasts. Broadcasts by radio hams use the same type of modulation as the ISS: Narrowband Frequency Modulation (NFM).

On finding a radio ham broadcast, take note of the following:

1. Notice how the signal looks and compare it with the local commercial FM radio station.
2. Look at the bandwidth of the signal and experiment with the software controls, in particular the slider control on the HDSDR software. The bandwidth needs to be set manually on the HDSDR software. On SDRsharp software, just click on the NFM button.

1. Watch the waterfall display, seeing how the display changes depending on what is being broadcast. Music will follow a rhythm, but speech will show short, random bursts of signals on either side of the carrier wave. Before long, it will be possible to distinguish between music and speech by just looking at the display.
2. Try recording the audio. On SDRsharp, check the audio box and uncheck the baseline box, changing from 16-bit recording to 8-bit.
3. Try playing the recording back. On SDRsharp, remember that after using the software to play back audio, change back to the SDR driver before closing the software.

On HDSDR software, just click record to save the audio and play to select the WAVE file. A useful feature on HDSDR is the loop playback button. This plays the audio back over and over, which is useful when making changes.

Picking up ISS Communications

Picking up communications from the ISS is quite easy to do and is a good project to start with, as it gives the radio operator a chance to learn some practical skills. One important skill is to understand how the Doppler shift affects the incoming signal and how to correct for the drift in frequency. As the ISS travels overhead, the radio operator will need to change the frequency to allow for the Doppler shift in the signal. The shift is predictable, as the ISS is traveling at the same velocity. The shift in frequency can be as much as a few hundred Hertz. These skills and others gained from this project will prove invaluable for use with later projects.

The first thing that needs to be done is to find out where the ISS is in relation to the observer. This is easy to do—just search on the internet for "Where is the ISS now?" A number of search results will be found. Click on any of these sites and the position of the ISS will be shown in near-real time. Figure 15.1 shows the ESA webpage. Here is a link to the website: https://isstracker.spaceflight.esa.int/.

Some websites will allow the observer to input their location, and the website will calculate dates and times when the ISS will be visible from the observer's location over the next few days. A good site for this is "Spot the Station." Here is the link to the homepage: https://spothestation.nasa.gov/ If you sign up, you will receive an email or text alert when the space station

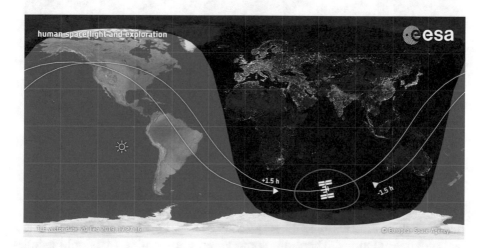

Fig. 15.1 Image showing the ESA webpage for tracking the ISS

is flying overhead. As with most things nowadays, apps have been developed and can be downloaded for use with smartphones. Here are two of them: the Google Play app *ISS Detector: See the Space Station*, and the Apple app *ISS Spotter*. These apps allow the user to know where the ISS is at any time. They haven't been used by the author, so no comment can be made of how good or bad they are.

Once this information is known, check that the observing site offers a suitable view of where the ISS will be seen to travel across the sky and that the view is not blocked by tall buildings or hills that are likely to hinder the signal. Try and avoid areas with high levels of interference such as industrial zones, especially where electric motors are used or there are overhead power lines.

With optical astronomy, the higher an object appears in the sky from the observing site, the better; this is also true with radio astronomy. The ISS website will give the height at which the ISS will be visible from the observing site and the approximate length of time it will be visible for. A good time to start is when the ISS appears 20° or more above the horizon as seen from the observing site. Below 20°, the signal will be traveling through the thickest part of the atmosphere and can become degraded. Additionally, the antenna will pick up unwanted signals and interference.

The radio signal from the ISS is a reasonably powerful one, so in theory it should be easily received.

Antennas to Pick Up the ISS

For the best chance of success, the antenna needs to be outside in the open, if possible away from strong radio sources. Some amateurs have received the ISS using the antenna that comes with the SDR receiver, although with some SDR packages, all that is provided is a small wire whip antenna. These wire whip antennas are fine to pick up local radio stations, and if very lucky, they may pick up the ISS. Practically speaking though, the antennas that come with some SDR packages are of poor quality, the small wire whip antenna may not give a good enough gain to receive the ISS.

Some observers have used the short, stubby, rubber-coated antennas sometimes referred to as *rubber ducks* pictured in Fig. 15.2. These rubber duck antennas are flexible and very robust. They deliver good reception as walky-talky antennas. There are some video clips online of people using them to pick up the ISS. Sadly, these don't seem to work here in the UK in a suburban surrounding. Maybe if they were tried in the middle of a large open space without surrounding buildings, they may work.

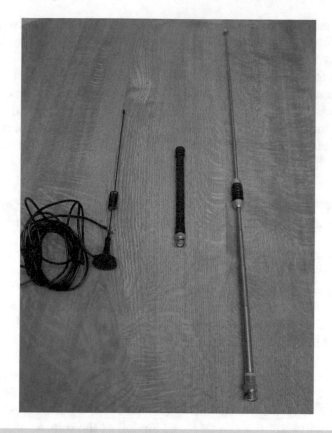

Fig. 15.2 Image showing three types of antennas. On the left the small wire antenna that came with the SDR. At center is the rubber duck antenna. Right is a good quality ¼ wavelength whip antenna, fully extended

Through experience, it has been found that a good quality ¼ wavelength whip antenna, as pictured on the right-hand side in Fig. 15.2, is a better choice. (A ¼ wavelength antenna is the length of a ¼ of the length of the radio wave it is designed to pick up. For example, a frequency of 145 MHz would have a wavelength of 2 m, so a ¼ wavelength antenna would measure 0.5 m). These ¼ wavelength antennas work really well over a good range of frequencies and can be easily extended. They cost somewhere in the region of $25 (£15) and can be ordered online. They are sometimes used with handheld scanners and can be picked up at ham radio shows.

Using one of these antennas is ideal, as it makes it possible to pick up the ISS day or night when the ISS is in the sky over the observer. If the observer feels the whip antenna does not give enough gain, try using a 3- or 4-element

Yagi antenna. If a Yagi antenna is to be used, the help of an assistant will be needed. Fit the antenna onto a wooden handle, such as a broom handle, in such a way that the broom handle can be held like a rifle. When the ISS is spotted in the sky, the assistant should look down the length of the handle like a rifle sight and line up the Yagi antenna with the ISS. This will allow you to track the ISS as it travels across the sky and keep it within the beam of the antenna. The main drawback of doing it this way is that it can only be done between dusk and dawn, as the ISS must be visible for it to be tracked with the antenna.

What to Expect

The ISS is only visible for around a maximum of 6 min at a time as it travels from horizon to horizon, as seen by an observer on the ground. The time will change depending on the altitude at which it is seen: lower altitude mean shorter time, while higher altitudes mean greater time. Six minutes isn't long, therefore only snippets of conversation can be heard.

The downlink frequency to receive the ISS is around 145.800 MHz. Try looking for the signal between 145.500 MHz and 146.100 MHz, as this will allow for the Doppler shift. If no signal is received, widen the search.

The ISS signal is modulated using Narrow Frequency Modulation (NFM). When looking for the ISS signal, have the NFM box checked on the SDRsharp software; this can be found top left of the screen.

On HDSDR software, click on the FM modulation button and then use the slider control, near the 12 blue control buttons, to change the bandwidth.

If the signal cannot be seen, try using the "zoom" function found on the top right of the screen on SDRsharp, and on HDSDR, the horizontal slider marked "spectrum zoom." Using the zoom control on SDRsharp, the resolution of the screen will need to be increased. This is done by clicking on the FFT control and increasing the screen resolution by two or three steps. Using the zoom control, try looking closely around the 145.800 MHz mark. Search for a thin peak standing above the background noise level. The signal looks just like a radio ham broadcast but could be weaker. Watch this peak for a few seconds. If it starts to drift in frequency from left to right on the screen, click on this peak and listen for dialogue. A tip if using HDSDR software: if the audio sounds a little garbled, instead of altering the frequency, try moving the bandwidth slider, as this may clear up the audio.

As the ISS travels overhead, be ready to change the frequency to continue picking up the signal. The frequency can change by several hundred Hertz. It is surprising how quickly it changes, especially when using the zoom feature on the software, as this magnifies the change.

On your first attempt at picking up the ISS, concentrate on finding the ISS signal and aiming for a few seconds of signal or dialogue. Watch how the frequency changes and drifts as the ISS travels overhead. Try and keep up with the change. Once this has been mastered, record the signal for examination later. On SDRsharp, check the record audio box and uncheck the baseline box, and also change the setting from 16-bit to 8-bit. When recording on the HDSDR software, just click on the record button. Both record in WAVE file format by default.

Processing the Recording

If the received signal is of good quality, processing may not be needed, but if the signal is noisy, a little processing may help clean it up. It's good practice to make a copy of any recording before processing. Then if the processing goes badly wrong, a backup is always available.

Some computers have a built-in graphic equalizer. Try passing the recorded signal through simple filters on a graphic equalizer to boost wanted frequencies and reduce noise. If the computer doesn't have a graphic equalizer, a good piece of software is "Ocenaudio." Here is a link to the website: www.ocenaudio.com. The software is free to download and easy to use. Ocenaudio comes with an 11- and 31-channel graphic equalizer and a number of useful filters for enhancing a recording. Looking at Fig. 15.3, the graphic equalizer can be seen in the centre.

Note that all the slider controls are in the middle and set at zero. In this position, if an audio recording is played, it will be played as it was recorded with no processing. If one or more of the sliders are slid downwards to a negative value, this will suppress those frequencies within the audio as it's played back. If the sliders are slid upwards to a positive value, those frequencies will be enhanced when the audio is played back. This can be useful at filtering out an astronaut's voice from any interference within the recording.

There are a number of useful filters, including low- and high-pass filters and band-pass and band-stop filters. Each filter can have its value changed, then applied to the audio recording as it is played back. The low- and high-pass filters can be useful for blocking either low or high frequency interfer-

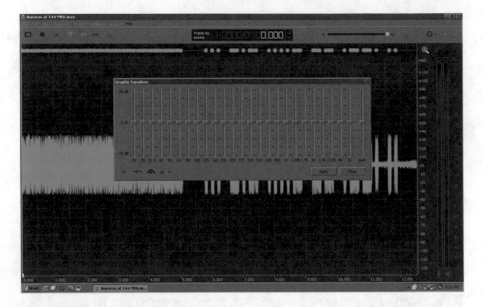

Fig. 15.3 Image showing Ocenaudio software with 31-channel graphic equalizer

ence. The band-pass filter and the band-stop filter will only allow the frequencies set by the operator to pass. This is extremely useful if trying to find a faint signal.

These filters will be covered in detail in Chap. 16. Experiment with each one to see how each one changes the audio output. If the last filter makes thing worse, just click the "edit" button top left and the last processing effect will be removed. A really useful effect is the amplitude filter, as it boosts the signal and can be used to tease out weak signals. Always remember to make a backup copy of the original recording.

The Best Chance to Hear the ISS

Check the ISS website to find out which astronauts are on the ISS at the time. If the signal received is of good enough quality, it maybe possible to recognize which astronaut is speaking. Sometimes the astronauts have a Question & Answer session with schools and other educational bodies, where students ask the astronauts on board a wide range questions. Details of these upcoming events can be found the ISS website.

Here are some useful links:

- https://www.nasa.gov/mission_pages/station/main/index.html
- https://www.nasa.gov/mutimedia/nasatv/schedule.html

For a new.sletter, go to:
- https://lp.constantcontact.com/su/hHN32CZ/nasagov.

Here are a couple of links to videos of amateurs picking up the ISS:

- https://www.youtube.com/watch?v=iy0GxzvQrqe
- https://www.youtube.com/watch?v=H45dNxZ7HGA

Chapter 16

Radar Detection of Meteors and the International Space Station

Meteoroids, Meteors, Meteorites and Micrometeorites

Debris left over by the tails of comets is stuck in orbit around the Sun. Mixed in with these cometary remains are small particles that have come from other sources as well, such as the collision between asteroids within the asteroid belt or the rubble left over after the Solar System came into being. These small particles are drawn towards the inner Solar System by the Sun's gravity and can get trapped within the debris left by comets. Together, this debris and particles are known as *meteoroids* and have been floating in space for many years. They are gravitationally attracted to each other and build up to form dense regions.

As the Earth makes its annual orbit around the Sun, it encounters these meteoroids. Once these meteoroids enter the Earth's atmosphere, their name changes and they become *meteors*. These meteors are traveling at high velocities between 8 and 85 km/s (5–50 miles/s). They burn up in the atmosphere and leave a streak of light in the sky. Not all meteors are made of stone. Some are made of iron, nickel or a mixture of the two metals. There are also stony iron meteors.

It is possible to see approximately ten meteor trails every clear night. Or, there can be a large number of meteoroids entering the atmosphere in a short time. Over the space of a few nights, this is what is termed a *meteor*

S. Arnold, *Radio and Radar Astronomy Projects for Beginners*, The Patrick Moore Practical Astronomy Series, https://doi.org/10.1007/978-3-030-54906-0_16

shower. On the nights of a meteor shower, all the meteor trails seem to originate from a single point, called the *radiant*. There are several well-known annual meteor showers. These are not named after the comet that produced the debris, but after the constellation that contains the radiant, for example the Leonid meteor shower. The radiant for the Leonid meteor shower is the sickle shape (or backward-facing question mark) in the constellation of Leo the lion.

When a meteor or small asteroid actually hits the Earth's surface, it is no longer called a meteor; it becomes a meteorite. These are larger pieces of debris that fall from the sky, not linked to cometary debris, and can be anything from a stray piece of the asteroid belt to pieces of the planet Mars. It is estimated that the Earth picks up approximately 20 ton of debris from space every day! This amount comes mostly from smaller micrometeorites—microscopic dust particles that, instead of burning up in the atmosphere like meteors, are slowed due to their very small size and float down from the sky unnoticed, similar to sand or volcanic ash.

Since most of the Earth is covered in water, many meteorites will be lost in the Earth's oceans. If a large meteorite hits the ground, it can cause a crater, such as the iron-nickel-type meteorite that created the Meteor Crater in Arizona, the United States. Samples of the original meteorite have been found within the ejector surrounding this huge crater; it is estimated that the meteorite was around 45 m (148 ft) in diameter, with most of the original meteorite vaporized on impact.

Since the invention of satellite imaging, other large craters have been found in different parts of the world, but because of the weathering caused by the Earth's atmosphere, these craters went unnoticed until they were imaged from Earth orbit. Meteorites can be found in most of the deserts around the world, as they are less likely to be disturbed by human activity. In addition, most deserts have low annual rainfall, so any meteorite will not have suffered from too much weathering before it is found. An excellent place to find meteorites is Antarctica. As they travel through the Earth's atmosphere, they get very hot from the air's friction acting as a brake to slow them down. The outer surface becomes a matte black in color as the heating from the atmosphere chars the meteorite's surface. Being covered in ice, Antarctica's landscape is totally white, and the meteorites are easily contrasted, making them very easy to see on the ground. The extremely cold temperatures and dry atmosphere of Antarctica also help to preserve the meteorites. NASA funds regular expeditions to Antarctica to search for meteorites; these have been very successful.

Here is a quick and easy experiment to try. Get hold of two neodymium magnets. These can be ordered online for around $5 (£3) or salvaged from

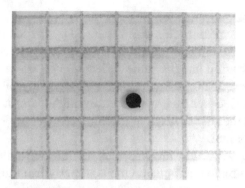

Fig. 16.1 Micrometeorite found using a magnet

an old computer hard drive. Keep them away from sensitive electrical items like cellphones, computers and the magnetic strip on the back of credit cards. Place the magnets in a plastic bag, then pop them in a container under a drainpipe and wait until it rains. When it stops raining, go outside and remove the magnets from the container. Examine them closely: with any luck, there may be some micrometeorites that have washed off the roof and gotten stuck to the magnets. They are small and can easily be missed. Figure 16.1 shows a micrometeorite found in the way as described above.

To give an indication of scale, in Fig. 16.1, the squares on the graph paper are 1 mm (0.04 in.) square. Note that this method only works on metal micrometeorites, but it is surprisingly effective.

Radar Detection of Meteors

The meteors that we are concerned with in radar detection are the small-sized meteors that burn up in the Earth's atmosphere at an altitude of approximately 80–100 km (49–62 miles).

The first radar observations of meteors were carried out by J.S. Hey (whom we met in Chap. 1) in the mid to late 1940s, just after World War II. Hey and his colleagues, working with army radar, were following up on reports from radar operators about the existence of short-duration echoes that they had heard. A number of theories were put forward to explain what these unknown echoes may be, such as attempts at jamming incoming radar signals, cosmic rays or the echoes from meteors high up in the Earth's atmosphere. Hey and his colleagues proved that the echoes did in fact come from meteors entering the Earth's atmosphere.

Hey found that the reflection of the radar signal was due mostly to the trail left behind the meteor. This trail is caused by the localized heating and ionization of the atoms within the atmosphere. The reflective properties of the ionized meteor trail had the same reflective properties on a radar signal as if it had been reflected off of a metal object. Hey also found that the number of meteor echoes was considerably higher on the days of meteor showers. He was the first to observe a meteor shower during daylight hours using radar equipment.

There are two ways that meteors can be detected entering the Earth's atmosphere using radio equipment. One is directly from the meteor itself, if it is large enough. The second and more practical way is by the trail that the meteor leaves in the sky, as this offers a far larger target.

Meteors produce a streak of light as they travel across the sky. These trails can last from a few seconds to nearly a minute, depending on the size and angle at which the meteor enters the atmosphere. As explained before, the trails are caused by friction as they enter the Earth's atmosphere at very high velocities. The heat generated by friction heats the atoms within the atmosphere and causes them to become ionized. In an ionized state, the atoms become highly reflective to radio waves and can bounce a radio signal.

Radar systems are expensive, so we need to "borrow" a radio signal to bounce it off the meteor's trail and detect them. We can use the carrier wave from a television transmitter, a radio station transmitter or space radar. The transmitter needs to be situated below the local horizon and not directly detectable from the observing site. The receiver is set to the carrier wave frequency of this transmitter or a little off the exact frequency to allow for the effect of the Doppler shift on the returning signals (more about this later). The theory is that the carrier wave from the transmitter is transmitted in every direction, including straight up into the sky; because the transmitter is below the local horizon, the signal cannot be picked up directly as the Earth is in the way. As a meteor enters the Earth's atmosphere and produces an ionized trail, the carrier wave signal hits the ionized trial or the meteor itself and is reflected back down to the Earth's surface, hopefully in the direction of the antenna and receiver. For a short time a signal would be received, until the ionization in the meteor trail is lost or the meteor burns up. See Fig. 16.2 for a simplified drawing of how radar meteor detection works.

The United States use powerful space radar to monitor satellites and other objects in Earth orbit. Radar meteor observers in the United States use the signal from this space radar for meteor detection. There are transmitters in Belgium, France and Sweden that are used by radar observers in the UK

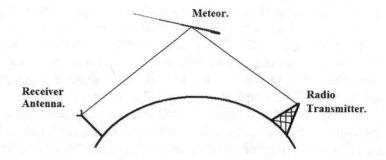

Fig. 16.2 Image showing the theory of radar meteor detection

and Europe. The French space radar is a good one to use if living in Europe and transmits at a frequency of 143.050 MHz, which is ideally suited for radar meteor detection.

Here is a link to a list of worldwide high frequency radio beacons: https://iaruhfbeacons.files.wordpress.com/2019/07/worldwide-list-of-hf-beacons-july-2019.pdf.

Find a transmitter that transmits at a frequency within the region of 50–200 MHz, as this range is highly suitable for radar meteor detection. If a transmitter within this range cannot be found, but one that transmits at a higher or lower frequency can be, as long as the receiver is capable of tuning to the frequency from this transmitter, it is worth giving it a try. If two suitable transmitters can be found, it is of great advantage, especially if the transmitters are in opposite directions from the observer's location, i.e. north/south or east/west. In this case, the transmitter that best suits the direction of the meteor radiant can be selected, or you can even use both at the same time, although this would require more equipment.

Radar detection of meteors offers some advantages over visual observations, including that observations can be made during the day and they are unaffected by clouds or light pollution.

The Best Times to Listen

It is possible to hear the echoes from meteors all year 'round, but there are certain days and nights when the number of meteor echoes can increase greatly—these are the annual meteor showers. During such periods, it may be possible to pick up the echoes of several hundred meteors per hour.

Table 16.1 gives some of the better known and most reliable showers with a brief description of what to expect. Please note the zenith hourly rates on the table are an indication of the expected rates and are an average taken from many years of study by dedicated meteor observers, both visual and radar. These can vary wildly and are not set in stone. It's possible to have half an hour of very high activity and nothing more than a few meteor echoes per hour for the rest of the night. There are other less well known meteor showers with maximum zenith hourly rates of around three per hour.

All the dates used with the table can vary a day or so due to natural variations in the width of the debris path and the Earth's orbit.

Any good monthly astronomy publications will have details of forthcoming meteor showers, or check online using any good astronomy websites. The dates given in astronomy publications can give the duration for a particular

Table 16.1 Well-Known Meteor Showers

Shower name and radiant	Month	Date of maximum	Approx. zenith hourly rates	Description
Quadratics	January	Around the 4th	50–100	Narrow maximum
Lyrids	April	Around the 21st	10–15	Moderate display, but can put on a surprise show
ή Aquarids	May	Around the 4th	20–50	Not easily seen from high northern hemisphere latitudes. Best seen in southern hemisphere
δ Aquarids	July	Around the 28th	15–25	This shower has a double radiant
Perseids	August	The glorious 12th	50–80	Rich annual shower; persistent trails
Orionids	October	Around the 21st	20–30	Fast-moving, persistent trails
South Taurus	November	Around the 3rd	10–15	Slow-moving meteors
Leonids	November	Around the 16th	15<	Very fast meteors; persistent trails. Can put on a good show from time to time. Is thought to follow a 33-year cycle of activity. Due again 2032?
Geminids	December	Around the 13th	50–100	Richest annual shower, usually an excellent display. Bright trails
Ursids	December	Around the 22nd	10–20	Can sometimes put on a good show

meteor shower, which can be anything up to 2 weeks. Along with this information will be dates and times where the maximum activity is expected to be.

The general rule is that meteors are more common, both visually and by radar detection, after midnight local time. This is because the Earth's rotation at this time has started to turn the Earth directly into the stream of debris, as seen from the local location. This effect is similar to the expression, "you only get squashed bugs on the windshield of a car, never the rear window." So, the best window is from midnight to 6 am local time.

There are limitations with radar detection of meteors, just as there are when observing them visually. Some meteor showers will be picked up better than others, depending on the height of the radiant as seen from the observing site. This is why it is a good idea to have more than one transmitter to tune into. It is a case of trial and error, but the effort is worth it. Monitor a year's worth of meteor showers; this will give you an idea of the effectiveness of the equipment and the best meteor showers to observe from your location.

Setting the Equipment

The SDR receiver needs to be set to the frequency of the transmitter being used and in the upper side band (USB) mode. In this mode, only the high frequency section of the signal is used. The frequency needs to be set slightly less than the frequency of the transmitter being used to allow for the Doppler shift within the received signal. For example, here in the UK, using the transmitter in Graves, France, which transmits at 143.050 MHz, the SDR receiver would be set to 143.048 MHz.

The beauty about using the SDR software is the visual interpretation you get for the signal. Using the zoom feature on the software, the USB bandwidth will be shown. As the meteor echoes are received, aim to get the echoes within the center of the USB bandwidth. A little fine tuning of the frequency up or down may be needed.

When the SDR is set in USB or LSB, the software will create a tone every time a signal is received. Thus, every time a meteor echo is detected, a "phwoooo" sound will be heard.

At first, each of these may all sound the same, but with practice, it will be noticed that each one has a subtle difference in character. This is because meteors enter the atmosphere at different angles, travel in different directions and have different velocities.

The sound of the echo that is received will be Doppler shifted. An echo's pitch and tone can and does change. For example, think of the whistles that

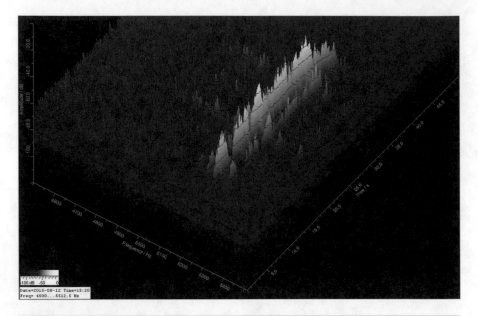

Fig. 16.3 A screenshot showing a 3D plot of a meteor echo with no shift in frequency

have a plunger in the base. Blowing the whistle and moving the plunger up and down changes the pitch of the tone. This gives the observer some idea in which direction and at what speed the meteor entered the atmosphere. The two screenshots in Figs. 16.3 and 16.4 show this quite well. The first figure is a three-dimensional screenshot using Spectrum lab software. Frequency is plotted along the bottom axis with time on the right-hand axis. It can be seen that the frequency stays the same over the duration of the meteor echo. The second figure shows the opposite effect, a rapid change in frequency over a short period of time.

In Fig. 16.3, there is almost no change in frequency, so the meteor that produced this echo was probably traveling at a right angle to the observer. In the second image, there is a large change in frequency over a short time, which means the meteor that produced this echo was traveling towards the observer. The three-dimensional display on processing software is very useful for showing this effect.

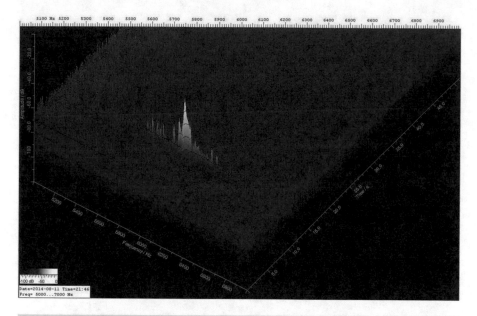

Fig. 16.4 A screenshot showing a 3D plot of a meteor echo with a rapid change in frequency over a very short period of time

How to Tell a Real Meteor Echo from Other Echoes

The following should help you distinguish the differences between the echoes of various objects.

Satellites

Satellites produce small echoes of very short duration in the order of a fraction of a second. This is because unlike a meteor, they produce no ionized trail within the atmosphere.

Aircraft

It is possible to pick up aircraft. Unlike satellites, they produce longer echoes with regular spacing. This is due to the speed that the aircraft is traveling. Five hundred miles per hour for a commercial airliner may seem fast, but compared with a satellite or meteor, it's slow. Aircraft also follow a specific course, so look to see if there is a pattern in the echoes.

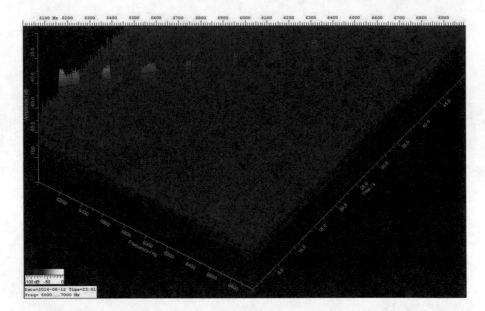

Fig. 16.5 The echoes from the ISS as it traveled over France and onto Egypt, picked up in the UK

Picking up the ISS

The ISS is easy to pick up. Using the ISS tracker, you should know exactly where it is at any time. Picking up the ISS is also an ideal way to test the limits of your equipment. By using echoes from the ISS, you can work out the lowest and highest latitude that the equipment remains effective. Figure 16.5 was taken from the UK. The ISS was picked up traveling across southern France on its way to Egypt. It looks like a large aircraft, but traveling at the speed of a satellite, about 27,842 kmph (17,300 mph).

Frequency is plotted across the bottom, and time on the right-hand axis. Look at the left-hand side; traveling across to the right are short-duration echoes. The duration of these echoes matches exactly with the space radar in Graves, France. The echoes are of uniform shape, size and duration. The change in frequency against time is a straight line showing a constant Doppler shift—thus, the object is following a predetermined course, unlike meteors, which follow random courses. Looking closely at the previous screenshots of the meteor echoes, you will see no uniformity to them. This is the key to identifying meteor echoes from artificial objects.

What Type of Antenna to Use and Why

In radar meteor detection, if we choose an antenna that is highly directional, we can only observe that part of the sky which the antenna is pointing at, so meteors that appear in other parts of the sky will be missed. If we choose an antenna that has very little directionality associated with it, like a single dipole antenna, we cover a larger area of the sky, but the gain in any particular direction won't be as great.

We have to balance gain against directionality. This depends on the observing site and proximity to the radio transmitter available for use. A general rule is: use the antenna that has the least directional quality that will work. If good meteor echoes are received with a single dipole, great—if not, try a two-element Yagi. If this doesn't produce good echoes, try a three-element Yagi, but don't go any more than a four-element Yagi, as this is narrowing the field of view too much. A single dipole, a three-element Yagi and a four-element Yagi will work, but the three-element Yagi delivers strong echoes with a good field of view.

Meteor trails occur at a height of approximately 70–80 km (43–49 miles) in altitude, but the height at which the echoes will be received above the observer's local horizon will depend on how far away the transmitter is located. The further away it is, the lower to the horizon the echoes will be picked up. An approximation can be worked out mathematically if the distance from the observing site to the transmitter is known, but a more accurate way is to use echoes from the ISS. As the ISS orbits, it travels over different parts of the Earth and at different latitudes as seen from an observer on the ground. As the ISS travels overhead, take note of the lowest and highest latitudes that echoes can be received. By doing this over time, you can work out an accurate field of view of your equipment.

The height of the antenna doesn't seem to be that crucial. A lot depends on the local horizon from the viewing site. Find a transmitter that isn't blocked by tall buildings or hills and offers a good view of the horizon. For the author, working inland in the UK with a reasonably clear southern horizon and using the Graves space radar in France, an antenna height of 3 m (10 ft) for a three-element Yagi antenna works well. Experiment with the height of the antenna until happy with the echoes being received.

It helps to tilt the front of the antenna upwards by approximately 10° from the horizontal, as this seems to give stronger echoes and reduce interference.

With radar meteor detection, we are receiving a reflected radio signal, so the polarization of the original signal will probably have changed after its encounter with the meteor's ionized trial. For this reason, some radar meteor observers use a cross dipole antenna. This type of antenna has a vertical element and a horizontal element. The theory behind it is that whatever the polarization of the incoming signal, be it vertical, horizontal or even circular, the antenna should pick it up. In practice, the polarization of the antenna isn't that important, as the reflected signals come at a random polarization. The best thing is to try picking up meteor echoes with the antenna vertically, then try the same experiment with the antenna horizontally and see which one gives the strongest signal and the lowest level of interference.

Recording Software

Once the radar meteor detection equipment is working, the next logical step is to record audio. Both SDRsharp and HDSDR can do this as described earlier, and the audio can be processed later.

A quicker way is to use two computers and process audio while it's being recorded. This is done by using an AUX cable. Plug one end of the AUX cable into the headphone socket of the computer running the SDR software, then plug the other end into the inline connection on the recording computer. If the computer doesn't have an inline connection, the microphone connection can be used, but make sure any microphone boost control is switched OFF. Care must be taken when setting the input volume on the recording computer if you are using the microphone connection, as the volume can easily overwhelm the computer's sound card.

A good rule to follow is to set the input volume so it is no more than a third up the scale when recording the background "hiss." This will allow two thirds to be used for meteor echoes.

There are two programs highly suited for recording meteor echoes, which we have seen before in Chap. 11; these are Spectran and Spectrum Lab. Both programs are equally good for recording meteor echoes. Spectran is easier to set up for recording meteor echoes when first starting out, while Spectrum Lab has the added advantage of the three-dimensional display. Try both programs and see which is preferred.

One advantage of Spectran is that it has an easy to set audio filter. When this is used with radar meteor detection it filters out a lot of the unwanted signals and makes it easier to find the meteor echoes. The filter works by only recording the frequency band set by the user and ignoring the frequencies above and below it.

Here is how to set the audio filter on Spectran:

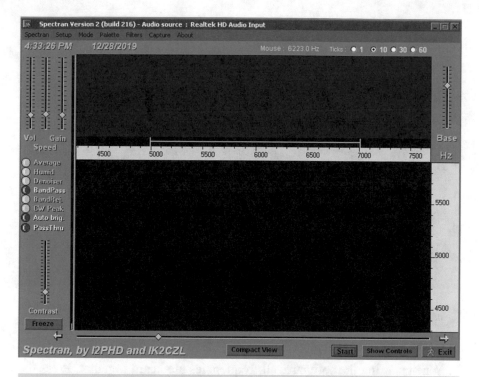

Fig. 16.6 Image showing the start-up screen of Spectran software

1. Open the software the screen, which should look like Fig. 16.6.
2. Click on the "setup" button top left. Scroll down to "recording source," then click on "record processed audio." The software will now only record the processed audio and ignore the input audio.
3. Start the software by pressing the "start" button at bottom right.
4. Watch the "waterfall" display and wait for an echo. Take note of the audio frequency of where the echo is on the display. On the computer used here, the frequency is centered around 6000 Hz. Set the filter bandwidth 1000 Hz either side of the echoes frequency.
5. Look at the line on Fig. 16.6 set between 5000 Hz and 7000 Hz. This is the bandwidth. This is easily set by putting the computer cursor on the lower frequency and clicking the left mouse button, then putting the cursor on the high frequency and clicking the right mouse button.
6. Press the "BandPass" button on the center left. This tells the software this is the frequency to record.

Fig. 16.7 Screenshot of Spectrum Lab software in 3D mode

7. When starting the software, all frequencies are heard together. Press the "PassThru" button. This will only allow through the frequencies within the bandpass filter; all other frequencies will be ignored.
8. To start recording, press the "record" button at top center. Before the recording starts it will ask for a file name. This can be anything as long you can remember it; simply typing in the date is a good option as it makes the files easier to find.

Recording using Spectrum Lab is slightly different. Figure 16.7 shows a screenshot of the program in 3D mode. The controls are shown down the left-hand side.

Here is how to record on Spectrum Lab:

1. Start the software running.
2. Click on the "options" button top left and select "3D spectrum." This will change the display from 2D to 3D.
3. Click on the "freq" tab top left and change the min and max setting to match the display required. In the example above, 5000 Hz is the minimum and 7000 Hz is the maximum. This will set the values for the display.
4. Click on "file" button top left, scroll down to "audio files," then click on "save output in audio file."
5. The software will ask for a file name. Once the file name is entered, the software will record the audio.

Analyzing Software

Once the recording has been made of a day's worth of meteor echoes, a good idea is to save the meteor echoes and discard the parts of the recording with no meteor echoes. This condenses several hours of recording into a few minutes containing the best meteor echoes. A good way of doing this is to use editing software. There is a program ideally suited for the purpose of analyzing meteor echoes. The software is called "Audacity" and can be downloaded at www.audacity.en.softonic.com/download. Audacity is not overcomplicated and is very intuitive and easy to use. Figure 16.8 shows a screenshot of the Audacity software.

Top left are the play, pause, fast forward and rewind controls; top centre are the cut and paste tools. If unsure of any of the program's functions, click on the "Help" tab at the top left, and then click on contents. This reveals a full list of functions and instructions on how to operate the program.

The display of the sound file on the screen of Audacity can be easily changed. Zoom in or out to give a more detailed view of the sound file. This is a good feature when working with a sound file of many hours. Figure 16.8 shows a graphic representation of a meteor echo recording. This one lasted for several seconds (meteor echoes can last from a few seconds up to a minute).

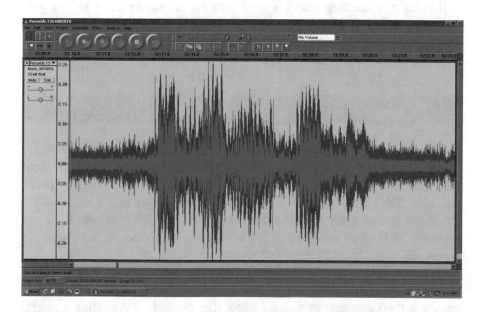

Fig. 16.8 Screenshot of the Audacity software showing a meteor echo

It is possible to receive some unwanted signals and/or interference with the meteor echoes. These unwanted signals have to be removed from the recording and discarded using the cutting tool. A good feature of Audacity is that if a potential meteor echo is seen, a slider can be dragged from the left-hand side and placed on the graph a few seconds before the suspected echo. Then you can press the play button on the control panel above. This will play the sound file from that specific point to help confirm whether it is a meteor echo or interference. This is useful while learning what to look for within the graph.

When a number of meteor echoes have been collected, they can easily be cut and pasted together to form a shortened version of the original recording. A tip: when doing this, allow a few seconds on either side of each meteor echo when editing. Two or three seconds will be enough—if they are placed too close together, when played back it will sound something like a firecracker. Any edited recording can be exported as a WAVE file and saved as a separate sound file. A word of caution when using Audacity software: after processing, save as a WAVE file and *not* as an Audacity file, as no other software will open an Audacity file. The edited recording can now be played back through either Spectrum Lab or Spectran. Observe how each of the echoes causes the frequency to drift higher or lower on the spectrogram window. Spectrum Lab is especially good at this because it has the three-dimensional display.

If a really interesting effect is seen, like a multiple signal echo or a particularly strong echo, both Spectrum Lab and Spectran have the ability to instantly save the display. Using this feature, it is possible to build up a selection of interesting meteor echo effects.

Automatic Meteor Counting Software

The theory behind meteor counting software is simple: it automatically counts the number of meteors and allows the user to keep a record of meteor numbers over several days. There are several versions of such software available. They all work in a similar manner—the user sets a limit within the software, and every time the limit is broken, the software counts this as a meteor. The theory seems perfectly sound, but in practice, it is an absolute nightmare to get running with any accuracy.

If the limit is set too high, the software misses some of the smaller meteor echoes. If the limit is set too low, every time a bit of interference is picked up, the software counts this as a meteor (for example, every time a neighbor's heating thermostat switches on and off this, will be included in the counter). Unfortunately, after years of attempts to get this software working with any accuracy in a sub-urban environment, the "just right" setting has still proven to be incredibly illusive.

Online Help and Other Useful Information

Podcast

A good first step would be to listen to the podcast by Steve Carter for the 365 days of astronomy project, as he wrote and recorded a great podcast on this very subject. Hear some excellent samples of the haunting noises made by meteor echoes; Carter also gives a brief description of how to do it. The podcast is titled "Meteor detection by radar for the amateur observer." Here is the link: www.cosmoquest.org/x/365daysofastronomy/?s=meteor+detection.

Websites

There are a number of useful websites where advice on radar meteor detection can be found. Here is a short list:

- Radio meteor detection: www.skyscan.ca/radio_meteor_detection.htm
- Meteor detection-NLO radio astronomy: www.meteorscan.com/
- American Meteor Society: www.amsmeteors.org/ams-programs/radio-observing/
- Detecting meteors using radio: www.meteorwatch.org/science
- Radio meteor listening: www.spaceweather.com/glossary/forwardscatter.html
- United Kingdom Radio Astronomer Association (UKRAA): www.ukraa.com

The British Astronomical Association (BAA) has recently started a radio astronomy section at www.britastro.org. The Society for Amateur Radio Astronomers (SARA), based in the United States, should be able to help with any enquiries concerning radio meteor detection at www.radio-astronomy.org.

Books

Norton, Richard and Chitwood, Lawrence. *Field Guide to Meteors and Meteorites*. Springer. 2008. This book has been very well thought out and put together. It covers meteors and meteorites in great detail, including the different types of meteorites, both stone and metallic, and the physical properties of each.

Chapter 17

Moon Bounce

Moon Bounce Experiment

This project came about while trying to get the radar detection of the aurora project to work. I needed to test the gain and beam width of two antennas to see which would be best suited for this project. The antennas in question were a four-element Yagi and a ten-element Yagi. What better way to test these things than to bounce a signal off the surface of the Moon and try to pick up the returning echo?

The four-element Yagi antenna didn't have much of a chance of pulling this off, but the ten-element Yagi proved its worth during this test. This experiment imitates Project Diana (see Chap. 1) carried out by the United States in the 1940s. Using a transmitter designed to transmit at a frequency of 111.5 MHz in 0.25-s pulses with a power output of 3000 W, scientists hoped to send a signal all the way to the Moon and back. The antenna could only be moved in azimuth, so scientists had to wait for the Moon to be in the right part of the sky and within the beam of the antenna.

The first successful detection of an echo from the Moon was by J.H. Dewitt and E. King Strodola in January 1946. The returning echoes were received approximately 2.5 s after transmission. Today, radio amateurs use the Moon to bounce their signals to other radio amateurs on the other side of the world. This is known as *Earth-Moon-Earth* (EME) transmissions.

S. Arnold, *Radio and Radar Astronomy Projects for Beginners*, The Patrick Moore Practical Astronomy Series, https://doi.org/10.1007/978-3-030-54906-0_17

There are a number of conditions for this type of transmission:

1. Both radio amateurs must be able to see the Moon from their location when the transmission is made and received.
2. The Moon as seen from Earth is only 0.5° in diameter, a relatively small target, so both radio amateurs must be able to move their antennas to keep them pointing at the Moon.
3. The Moon is on average 384,400 km (238,855 miles) from the Earth, so the transmitted signal must be of significant power to be able to reach the Moon and return back to the Earth.
4. The receiving antenna must have enough gain to pick up the returning signal, so a highly directional antenna is needed.

Finding a Transmitter to Do a Moon Bounce

The first thing to do is find a suitable transmitter. Here in the UK, the Graves space radar in France is ideal for this project, as it has a powerful signal. There are such space radars in the United States and other countries, but a powerful radio/TV transmitter will also do. Here is a link to a list of worldwide high frequency radio beacons: https://iaruhfbeacons.files.word-press.com/2019/07/worldwide-list-of-hf-beacons-july-2019.pdf.

The transmitter needs to be below the horizon, not directly detectable and preferably in a direction where the ecliptic is above it. The French space radar transmitter at Graves is due south as seen from the UK with the ecliptic due north of it, hence why it is ideally suited for this purpose.

Finding the Moon's Position

It is quite easy to find the Moons position, as any astronomy software will give the coordinates of the Moon. Spectran software has a feature to show the Moon's position. Open the software, click on the "mode" button on the top toolbar, scroll down and click on "show Moon;" up pops a window showing the Moon's position in azimuth and elevation.

One piece of software that as proved useful while doing this project is "Homeplanet."

This can be downloaded from www.fourmilab.ch/homeplanet/. The software is free to download. It shows the Moon as it crosses the Earth's surface, which is useful in seeing the Moon's position in relation to the observer's site and the transmitter. See Fig. 17.1.

Fig. 17.1 Image showing a screenshot of the Homeplanet software. The Moon is just left of centre, traveling over Africa

Tracking the Moon

As mentioned earlier, the Moon needs to be tracked to keep it within the beam of the antenna. This is easy using a telescope mount, either driven or undriven. The antenna will need to be fitted to the mount, which can be done by fitting the antenna to a block of soft wood, then gripping the wood block in the clamp on the mount. Using soft wood shouldn't damage the mount's clamp.

After polar aligning the mount, fit the antenna and check the mount's balance as shown in the operating instructions. A weight may need to be added to the other end of the wood block to balance the antenna. If an undriven mount is used, small corrections will need to be made from time to time to keep the Moon within the antenna beam.

A better method is to use a driven mount such as an EQ 3 mount. This type of mount will hold up to a 5.44-kg (12-pound) load and is more than capable of holding a 10-element Yagi antenna, although in windy conditions, care must be taken that the mount does not blow over. See Fig. 17.2.

Next, line up the antenna with the Moon, which can be done either by looking down the length of the antenna like a rifle site or using a small plastic red dot finder that can be temporarily fixed to the antenna using a rubber band. This finder, if fitted to the center of the antenna, can be left in place while the test is carried out without having any detrimental effects on the antenna.

Fig. 17.2 Image showing a 10-element Yagi mounted on an EQ 3 mount, used to receive returning Moon echoes

SDR Software Setting

Set the SDR software to the continuous wave (CW) mode, then set the frequency display to the frequency of the transmitter being used. The Moon at its closest distance from the Earth is 363,105 km (225,623 miles) and at its furthest is 405,696 km (252,088 miles). So, there will be a Doppler shift to the received echo, which will change the frequency slightly every time this is done.

Graves in France transmits at a frequency of 143.050 MHz. The best way to find the returning signal is to start looking at the frequency at which the transmitter is transmitting, then slowly move down the frequency scale until the signal is found. The returning echo will be a much weaker signal, so the zoom feature on the software will need to be used. Also increase the resolution on the display so the returning echo is not missed. See Fig. 17.3.

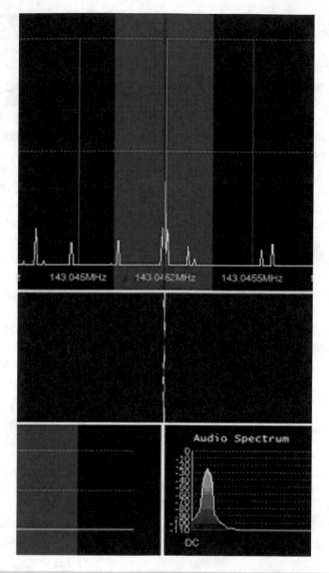

Fig. 17.3 Screenshot showing returning echoes from the Moon

Once a signal is found, you can perform a simple check to prove it is a genuine echo from the Moon. Watch the computer display and move the antenna so it is not pointing directly at the Moon; the signal should disappear. Return it back to the Moon, and the signal should reappear.

The best way to record the signal is using the SDR software as described in Chap. 14.

Working out the Beam Width of the Antenna

Once the returning echoes are being received, you can find the width of the antenna beam through the remote controls on the mount. Using the speed controls, speed up the mount in RA until the signal is lost, taking note of the RA coordinates. Then take the mount in the opposite direction until the signal is lost again, taking note of the RA coordinates. Then take one from the other. For example, if the signal is lost at 0000 hours in one direction and 0230 h in the other direction, the difference in the RA coordinates is equal to 2 h and 30 min. This will give a horizontal beam width of 37.5°.

The same thing can be done with the DEC coordinates using the speed controls on the mount. Increase the declination until the signal is lost, taking note of the DEC angle. Then lower the declination until it is lost again, taking note of the DEC angle. Take one angle from the other. For example if the signal is lost at a DEC angle of 63° at the highest point and a DEC angle of 37° at the lowest point, taking 63–37, we get a beam width of 26° in the vertical plane.

Another thing to take note of is the signal strength. Does it increase midway between the changes in RA and DEC when the antenna is at its maximum gain? It should, but the software may not show this change too well. The change may be heard as the signal strength varies.

Useful Links

Here is a link to a video clip showing SDR software used to pick up returning echoes from the Moon with the help of the Graves transmitter in France: https://www.youtube.com/watch?v=FHj2QxQ7mLE.

Here are links to video clips using two different types of scanners to pick up returning echoes from the Moon with the help of the Graves transmitter in France:

- https://www.youtube.com/watch?v=zv1UexVIDMI
- https://www.youtube.com/watch?v=xpcsRDI2bv4

Homeplanet software has a list of radio sources along with a planetarium. To find it, click on "Display" on the top toolbar, then "object catalogue," then "category," then "3C radio sources."

Chapter 18

Radar Detection of the Aurora

Myths, Legends and History

The aurora is as old as the Earth itself. There are many different myths and legends as to what the aurora could be. These include the spirits of dead warriors making their way back home or the souls of loved ones making their way to their final resting place. In Scotland, the aurora was thought to be fallen angels from heaven. There are also myths surrounding fertility—it was once thought that an auroral display signaled the ideal time to conceive, and that a baby born at the time of an auroral display would be lucky.

It is impossible to say who the first person was to look up at the sky and marvel at it. The first person to be credited with recognizing that the aurora was an astronomical event was the seventeenth century French priest, philosopher and astronomer, Pierre Gassendi. He thought that there was some sort of energy interacting within the Earth's atmosphere. What this energy was and how this interaction took place were to remain a mystery.

In the late 1890s into the early 1900s, the Norwegian scientist Kristian Birkeland investigated and became the first person to come up with a realistic scientific theory on the origin of the aurora. He suggested that the Sun could be sending out a constant stream of charged particles (the solar wind), which could be interacting with the Earth's magnetic field and causing the aurora. He built a number of small experimental vacuum tanks in which he managed to produce artificial auroras, which he called *auroral jars*. One of

S. Arnold, *Radio and Radar Astronomy Projects for Beginners*, The Patrick Moore Practical Astronomy Series, https://doi.org/10.1007/978-3-030-54906-0_18

his most famous experiments involved a large vacuum tank. In the centre of the tank he placed a hollow sphere to represent the Earth. Inside this sphere, he placed an electromagnet to simulate the Earth's magnetic field. At the other end of the vacuum tank was a metallic plate where Birkeland applied a high voltage somewhere in the range of 20,000 volts, which was to mimic the charged stream of particles coming from the Sun. When he switched on the electromagnet inside the hollow sphere, an aurora oval appeared around each of the electromagnetic poles.

This demonstrated the existence of the solar wind and its interaction with Earth's atmosphere to produce the aurora.

There is a famous image of Birkeland carrying out this very experiment. In it, he is wearing a Fez (a red headdress). It was rumored that he always wore the Fez, as he thought it protected his brain from any stray radiation coming from his experiments. Birkeland went on to carry out a number of other experiments in astronomical theories including the solar cycle and sunspot numbers, and made the suggestion that solar activity could have something to do with the Sun's magnetic field. He carried out work into the zodiacal light and was also fascinated by the rings of Saturn. Unfortunately Kristian Birkeland struggled with mental health problems, and this contributed to his early death.

Here is a link to an interesting article about Birkeland: https://www.dailyscandinavian.com/norwegian-scientist-solved-mysteries-spectacular-aurora-borealis/.

The Earth's Magnetic Field and Magnetosphere

The magnetic field of the Earth has already been covered in detail in Chap. 4. To recap how the aurora is formed, in times of high solar activity, the incoming pressure of the solar wind on the Sun side can increase greatly, causing deformation in the magnetosphere. The side away from the Sun can develop a long magnetic tail that can stretch many thousands of kilometers into space. As more and more energy from the solar wind hits the leading edge of the magnetosphere, the tension builds up in the long magnetic tail like an elastic band. At some point the elastic band snaps back, bringing charged particles careering along the field lines and funneling them down at the poles where they interact with the atoms in the atmosphere. This produces the aurora. See Fig. 18.1.

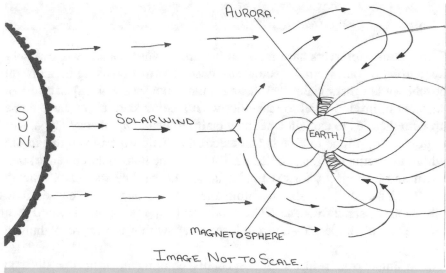

Fig. 18.1 The Earth's magnetosphere

Aurora Colors

As the solar particles hit molecules within the Earth's atmosphere, it releases a photon of energy. This photon is always of the same wavelength depending on which molecule in the atmosphere is hit. For example, the color red comes from collisions with molecules of nitrogen, and the color green comes from collisions with oxygen molecules. Coloring can vary from a strong, vivid color to a weak, wispy color.

Other colors have been reported, such as pale yellow, white, purple and blue. An explanation for these differences is that aurora observing is done at night in very low light, and so people will see the shades differently. This is known as the *Purkinje effect*. This effect causes some people to see some colors brighter than others even though they are the same brightness. It is well known to observers of variable stars, where the color red is often seen as fainter.

An excellent way to record auroras is to use a camera, as it has better color perception at lower light levels than the human eye and can give a truer color rendition.

Strange Sounds, Tastes, Smells and Feelings

Some aurora observers have reported hearing strange sounds while watching auroral displays, similar sounds to those of waves breaking on a distant beach, while others claim they hear a sound like voices whispering in an adjacent room, too quiet to make out what's being said. Many have come up with theories to explain what they claim to hear. One theory that seems to pop up from time to time is that electromagnetic waves from the aurora within the frequency range of 20 Hz–20 KHz (the human audio range) rain down and hit objects, for example a large, flat surface such as a fencing panel. This energy builds up within the fencing panel and causes it to vibrate at the same frequency. This vibration of the fence panel would then vibrate the air molecules and produce sound within the range of human hearing.

The simple truth is that without specialist equipment, auroras produce no sound within the frequency range of human hearing. Rather, it is thought that the brain subconsciously makes the observer think they are hearing these sounds. Some observers are so convinced they hear sounds that they go to great lengths to make audio recordings when observing auroral displays. As yet, none have been successful.

Aurora observers have also reported they can taste and smell the aurora in the air. They describe it as being similar to the taste and smell after a thunderstorm that has delivered a number of lightning discharges. There is no evidence for this. As before, it is thought that the brain knows the aurora is electrical in nature and subconsciously makes the observer believe they taste and smell the electricity in the air.

Still other observers claim to have strange feelings while watching an auroral display. They describe these feelings as having a static electric charge all over their body, or the hair on the back of their neck stands up. All things being equal, the simplest answer is usually the correct one. Probably, these things happen because the observer is feeling the cold or the excitement of seeing an auroral display in person.

Types of Auroras

There are several different classifications of auroras. Each one denotes a change in its appearance and behavior from the perspective of an observer on the ground. It is not uncommon for an aurora to change from one form to another as the night progresses.

From when it starts to get dark, an observer will see an early evening aurora known as a *glow aurora*. This is a slight brightening of the sky low down on the horizon, similar to the brightening of the sky an hour or so before dawn. This type of aurora can easily be mistaken for light pollution in built-up areas. In a location where there is little light pollution, hills or tall trees in the distance may appear to have a glow coming from behind them. The color of the aurora at this point will be very pale and will only show up on a photographic image, as the human eye isn't sensitive to the color just yet.

Later in the evening, an observer will see the glow climb higher in the sky and form a *homogeneous arc*, sometimes called a *quiet arc*. These arcs can be quite uniform in shape, not unlike a rainbow. The arc can stretch across the sky from east to west. Sometimes bands of light can be seen rising up from its top. The brightness within it can sometimes vary and even appear to flicker. Any movements within the arc are still quite slow. At this point, color within the aurora starts to be noticed. It may appear white, a very weak green or a mixture of the two.

Later still, the aurora starts to get more active and changes to a *rayed arc* or *rayed band aurora*. Now there are more bands of light rising up from the arc, which start to move more quickly as if dancing. The flickering in brightness of the arc gets stronger and travels the full length of the arc. The white color starts to disappear while the green color becomes more vivid, and new colors like pinks and reds are now seen. We also now start to see "to and fro" motion within the arc itself, like waves traveling down a length of rope. All changes within the aurora become greater and it starts to move faster.

At midnight local time, the aurora changes to the *corona aurora*. This is one of the most spectacular auroral displays. At this point, the aurora is directly overhead, as seen from the observer on the ground. Looking straight up, the observer will see an optical illusion, as if being surrounded by tall buildings that appear to converge at a single point high in the sky. These tall columns will often appear in strong, vivid greens, reds and purples. Everything is moving quickly, like a light show or fireworks display.

After midnight, the excitement of the corona aurora starts to die down, and the aurora turns into a *pulsating aurora*. At this time, it dissipates and becomes almost veil-like, losing its vivid colors and filling the sky. In places there will be cloud-like structures where the density is higher; these clouds start to pulsate at regular intervals, and here the colors are still quite strong.

As the night comes to an end, the thin veil of the aurora disappears, and then one by one the higher density cloud-like structures start to dissolve until they have all gone.

The Kp-Index

The aurora as seen from space appears to form a ring around both magnetic poles. An observer standing at either magnetic pole would see the aurora all the way around them—this ring is called the *auroral oval*. As the energy of the aurora increases, the auroral oval will expand and be pushed down towards the equator, or upwards in the case of the southern hemisphere. Auroras at the equator are rare but not unknown, especially at the times of solar maximum.

There are a number of different ways to measure auroral activity, but the Kp-index is an easy one to understand and use. The index was introduced by Julius Bartels in the 1930s ("K" taken from the German word "Kenizffer," meaning "characteristic digit"). It is a measure of the geomagnetic disturbance within the magnetic field of the Earth. The higher the number, the greater the disturbance and the further towards the equator the aurora oval is pushed, giving observers at the lower latitudes a chance to observe it.

The Kp-index works on a scale from 0 to 9 and is classified as:

- **Kp 0–1** no activity or very low activity
- **Kp 1–2** very low activity
- **Kp 2–3** low activity
- **Kp 3–4** medium activity
- **Kp 4–5** medium to high activity
- **Kp 5–6** high activity
- **Kp 6–7** very high activity
- **Kp 7–9** extremely high activity

The magnetic poles of the Earth are offset from the axis of rotation. This means the Kp-index-to-latitude relationship will change depending on where you are in the world. A quick search on the internet will show this relationship; just type in "Kp index + country." For example, this is the Kp-index as seen from the United Kingdom, where the change in Kp levels and latitude are as follows:

- **Kp 0–1** travels across the southern tip of Greenland
- **Kp 2–3** travels across the north and south of Iceland
- **Kp 3–4** travels across Norway
- **Kp 5** touches the tip of Scotland, the very top of the UK
- **Kp 6–7** travels through the middle of the UK
- **Kp 7–9** travels across the north and south of France

Geomagnetic Storm

A geomagnetic storm happens when there are major changes in the Earth's magnetic field. These types of storms can last from a few hours to a few days. They are caused when solar activity is high, such as solar maximum. Other types of solar activity that can cause them are coronal mass ejections, solar flares and high sun spot numbers. These types of storms can damage satellites, which control GPS systems, cellphone communications and radio and television communications. An alert is usually issued when Kp 5 is reached, with more severe warnings issued as the Kp number increases.

Detection of the Aurora

The beauty about radar detection of the aurora is that like radar meteor detection, it can be done through clouds and, best of all, during the day. The theory behind the following project is quite simple, although putting it into practice is far from it. There are a number of variables that will conspire to cause problems. Patience and perseverance are the keywords with this project.

When an aurora is present in the Earth's atmosphere, the energy imparted from the charged particles in the solar wind makes molecules in the atmosphere ionized, not unlike a meteor trail. In this ionized state, the atmosphere will reflect radio waves like a mirror reflects light. Using a reliable radio signal that can't normally be received by an observer, and also using the ionization of the atmosphere when an aurora is present as a reflector, the signal will be bounced in such a way that the observer will receive it.

First, find a suitable radio signal. This radio signal must be in the right location, transmit on the right frequency and have enough power to transmit over a great distance.

How to Find a Suitable Radio Transmitter

All over the world are high frequency radio beacons. Here is a link to a worldwide list: https://iaruhfbeacons.files.wordpress.com/2019/07/world-wide-list-of-hf-beacons-july-2019.pdf.

The beacon needs to be over the horizon from the observer and cannot normally be picked up with the receiver from the observing site. The right location is important: if the observer is in the northern hemisphere, the beacon should be to the north of the observer. The opposite applies in the southern hemisphere. The principles will be the same for any beacon, except for the different broadcast frequencies and location of the observer.

In this example, we shall discuss the beacon that is used for radar detection of the aurora in the United Kingdom. The beacon used in the United Kingdom is the SK4MPI beacon, located in Borlange, Sweden. It produces two powerful broad beams, one to the northeast and one to the northwest, and broadcasts on a frequency of 144.412 MHz. Its signal is broadcast in unmodulated continuous waveform and is of 1 min duration, then repeats.

With the SDR software set in continuous wave (CW) mode, an observer will hear a constant tone for 45 s coming from the continuous wave, followed by 15 s of dots and dashes spelling out the beacon call sign, SK4MPI, in Morse code. Here is a link to hear the SK4MPI beacon: https://www.youtube.com/watch?v=0jlEpmge3Hk.

Figure 18.2 is a visual representation of a recording made of the SK4MPI beacon. The left-hand side shows the continuous wave, while the right-hand side shows the call sign SK4MPI spelled out in Morse code.

It should be possible to pick up the reflected signal from the SK4PMI beacon with the right equipment when the Kp-index reaches Kp 4 and above. In practice, the observer should only hear the beacon at the times of auroral activity or for a second or two if the odd meteor passes through the beam. At all other times, just the hiss of white noise will be heard.

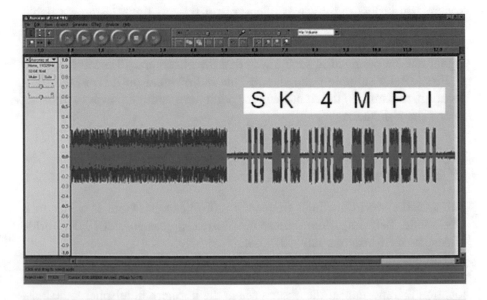

Fig. 18.2 Image showing the signal from the SK4MPI radio beacon

The reflective quality of a meteor's ionized trial can be compared to that of having a copper wire hung in the sky. When a meteor echo is heard it is clear and unmistakable, and any subtle changes within the signal can be heard. The ionization of the atmosphere caused by the aurora has poorer reflective qualities than that of a meteor echo; any reflected signal is always heard with background noise.

The frequency of meteor echoes falls within a narrowband width of a few tens of Hertz on either side of a central frequency. This makes them ideal for recording overnight and checking in the morning. When working on aurora detection, the drift in frequency can be large—a thousand Hertz on either side of the central frequency is not unheard of. The drift in frequency can happen quickly over a few seconds, and an observer must be ready to change the frequency up or down as the signal changes. This makes recording the aurora over long periods a bit of a nightmare.

Antennas for Use with Radar Detection of the Aurora

What we are trying to do is receive a signal that has traveled several hundred miles from its transmitter, hit a bit of ionized gas in the atmosphere, then gotten reflected and sent off in a different direction, only to travel several hundred miles more to the receiver. Just thinking about how far the signal has to travel, we can see how a directional antenna will be a must for the project.

In theory, in the United Kingdom the smallest recommended antenna for aurora detection is a four-element Yagi. This antenna has a wide pickup area, but the forward gain is low and only really good enough for when the auroral activity is very high. A better antenna for this project is a ten-element Yagi, as the forward gain is almost three times that of the four-element Yagi, although the pickup area is much smaller.

With these two antennas, it is a tradeoff between having a wide pickup area and a low forward gain, or a smaller pickup area and a higher forward gain. Using the ten-element Yagi with its smaller beam width and higher gain means the orientation of the antenna needs to be more accurate. These points proved to be the key to cracking this project after several years of trying.

Orientation of the Antenna

When setting up an antenna to pick up meteor echoes, it is fine to point the antenna directly in the direction of the radio transmitter, where any meteors or space stations traveling between the transmitter and the receiver will be picked up. This cannot be done with this current project, and orientation of the antenna is very important here. Just pointing the antenna north (south in the southern hemisphere) isn't accurate enough—it's knowing which north (or south) to use.

There are three kinds to consider:

1. **Grid north**: This north is used when drawing maps.
2. **True north**: This is the rotational north pole of the Earth that points to the North Star, Polaris.
3. **Magnetic north**: This is the one we are interested in. The magnetic north pole is offset from grid north. There are only a few degrees between them, but this can make the difference between receiving an echo or not.

Look at Fig. 18.3. This is how the orientation was worked out from a latitude of 53 degrees north in the UK.

Grid north is the vertical line. Offset to the left is the magnetic north. A line drawn at 90 degrees to the magnetic north line represents the aurora oval. The line on the left, drawn at an angle of 45 degrees, represents the northwest beam coming from the SK4MPI beacon. The dashed line represents a theoretical "normal," the line to separate the angle of incident from the angle of reflection. As can be seen, the antenna needs to be pointed towards the angle of reflection to give the best chance of receiving the reflected signal.

Observers in other parts of the world need to find out where the magnetic pole is in relation to their own location. The details of the high frequency radio beacon, such as the beacon's location, can be found by doing an online search. Where the beam from the radio beacon hits the Kp line, draw a line at 90 degrees to one of the Kp lines to represent the "normal." The angle between the "normal" and the beacon line will be the same as the reflected beam. This will show the observer in which way to point the antenna.

Antenna Height and Feed Cable

Place the antenna where there is a good view of the horizon, avoiding large obstacles like buildings and hills. The antenna needs to be at a minimum height of 3 m (10 ft) from the ground. If no signal is received, consider increasing the height; this may mean up to a height of 9 m (30 ft) or more.

Choose a coax cable with low loss characteristics for the feed, such as RG58, but make sure it's the RG58 military spec version. Another good choice is RG8 mini. Both are 50 Ω coax cables, but the RG8 mini has better loss characteristics.

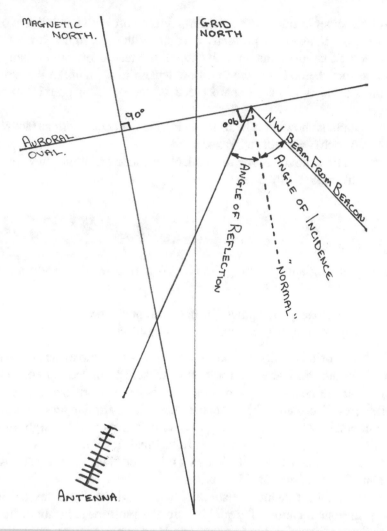

Fig. 18.3 Diagram showing antenna orientation

After cutting the feed cable to the correct length for the frequency of the transmitter, fit a number of ferrite collars on each end. Keep them as close to the end of the cable as possible, holding them in place with cable ties. When fitting the connections on the ends of the cable, try not to score the central conductor as you remove the insulation, as this can cause a weak point that could be damaged by movement from the wind when the antenna is in place.

With the outer braided wire shielding, don't worry too much if one or two pieces break off, as this is perfectly normal. With this project, it is better to use solderable connections on both ends of the feed cable rather than twist-on connectors. A good practice to follow when fitting an unfamiliar connection on a cable is to use a spare piece of the cable to practice fitting a connection.

Here is a link showing how to fit an F-type connector: https://www.youtube.com/watch?v=Bn30LdXiWx4.

Here is a link showing how to fit a BNC connector: http://www.youtube.com/watch?v=DksUM656s-m.

What to Do Next

Start by planning and checking the NOAA aurora website. Here are some links:

- www.swpc.noaa.gov/products/aurora-30-minute-forecast
- www.swpc.noaa.gov/products/aurora-3-day-forecast

On this website, details can be found in near-live time of what's happening with the aurora. There is a forecast of what is predicted to happen to the aurora over the next 30 min to 3 days in both the northern and southern hemispheres. There are other websites that offer a similar service and even send out emails and texts, but they may require a monthly charge. In some countries like the UK, if the observer inputs their location, they will receive a more personalized forecast. Search online for "aurora forecast." Again, some sites may charge for this service.

To have the best chance of success, wait until the Kp-index is high as seen from your location. Try picking up the radio beacon about an hour before the maximum is due to occur and keep trying until an hour after the maximum.

Set the SDR software to CW mode and set the frequency to that of the beacon's signal. Be sure to turn the AGC control off. As discussed previously, the AGC will try and keep the signal at a constant level, but the natural variation in the signal strength is what is needed to pick up the weaker aurora echoes.

Doppler shift within the signal will cause the frequency to change by as much as a 1000 Hz. Try sweeping the receiver's frequency up and down by as much as 1000 Hz on either side of the beacon's frequency. Listen very carefully for any signs of the beacon's signal while sweeping through the frequencies. It could be very subtle and hiding in the background noise.

There will be many false alarms. If in doubt, give it a few seconds before moving on. Speaking from experience, before celebrating, make sure the call sign of the beacon can be correctly identified to ensure the right beacon is being received (see Chap. 19).

If no signal is received, wait for the Kp-index to be higher. In the meantime, double check things like antenna orientation, beacon frequency and the settings on the receiver. If all these are right, increase the height of the antenna. If a signal is still not being received, consider upgrading the antenna to a 12-element Yagi or more. If a signal is being received, keep an eye on the frequency, as it can change quickly either up or down, and you will need to make quick changes to the software settings to keep receiving the beacon. Keep records of the results, as these are likely to change over the course of a year and of the solar cycle.

Try to ask yourself the following:

1. What is the lowest Kp number that it is possible to pick up from your location?
2. What is the earliest the beacon's signal can be received before the auroral activity is at its highest from your location?
3. Do the above points change from summer to winter?

Recording the Signal

Trying to record the echoes from the aurora is a bit of a nightmare for the following reasons:

1. Echoes from meteors give stronger signals, whereas echoes from the aurora are much weaker.
2. Echoes from meteors tend to hang around for at least a few seconds due to the ionized trail they leave, whereas echoes from the aurora come and go randomly.
3. The frequency range at which the echoes occur. Meteor echoes tend to stay within a range of 1000 Hz. Echoes from the aurora can change frequency by as much as 1000 Hz above and 1000 Hz below the beacon's frequency, and this change can happen in a matter of seconds.

The best way to record aurora echoes is to start the software recording and try to keep up with the constantly changing frequency while the software is recording. Afterwards, editing software such as Audacity can be used to isolate the echoes. As mentioned earlier, the signal changes in strength and frequency at random, so this can be a real challenge to do.

Chapter 19

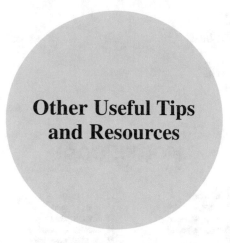

Other Useful Tips and Resources

Radio Beacon Recognition

The high frequency beacons used within this book all have their own unique call sign. The trick is to know the right beacon has been found. For example, if we use the SK4MPI Radio Beacon used to detect aurora in the UK, its signal consists of a 60-s repeating signal. The first 45 s are the constant beep of the continuous wave. Followed by 15 s spelling out the SK4MPI in the international Morse code. To use these beacons, some knowledge of the international Morse code will be needed.

The best way to learn the call signs is by sound, not by looking at a chart of the Morse code as shown in Fig. 19.1, which is included only as a reference guide.

When learning Morse code, the first thing is *not* to think of the letters and numbers as being made up of dots and dashes, but to think of the sound they make when sent. A dot sounds like a "dit" and a dash sounds like a "dah." For example, the letter "A" is made up of one dot and one dash, but when heard in Morse code, it would sound like "dit dah." The letter "B" is made up of one dash and three dots, which would sound like "dah dit dit dit."

© The Editor(s) (if applicable) and The Author(s), under exclusive license to Springer Nature Switzerland AG 2021
S. Arnold, *Radio and Radar Astronomy Projects for Beginners*, The Patrick Moore Practical Astronomy Series, https://doi.org/10.1007/978-3-030-54906-0_19

So, the SK4MPI beacon sounds like:

- S = "dit dit dit"
- K = "dah dit dah"
- 4 = "dit dit dit dit dah"
- M = "dah dah"
- P = "dit dah dah dit"
- I = "dit dit"

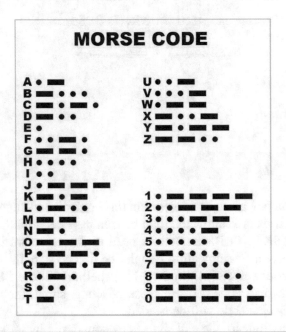

Fig. 19.1 The international Morse code (for reference only)

A good way to learn the sound of a beacon's call sign is to use software to mimic its call sign in Morse code and become familiar with how it sounds. A good piece of software for this is "Just learn Morse Code." Here is a link to download this free software: http://www.justlearn-morsecode.com/download.html. The software is easy to use. Just type the call sign of the beacon in plain text, and the software will play it in Morse code. This can then be recorded using the software and played back at anytime.

This all may sound confusing to start with, but by playing the recording several times a day, it is surprising how quickly it becomes familiar.

Building a Tesla Spirit Radio

The Tesla spirit radio from Chap. 1 is an easy project to build. The one thing to be careful with is the diode. This must be fitted the correct way around and it is easily damaged by heat, so care must be taken when soldering it in place. Use a pair of tweezers as a heat shunt to prevent damage when soldering.

Before building a Tesla spirit radio, check out these useful resources:

Here is a link to a PDF file showing the radio's parts list, circuit diagram, construction ideas plus loads more information: http://cdn.instructables. com/id>ORIG>FJY>ISIM.

All parts can be ordered separately online but it may be cheaper to buy a kit, which can be purchased from the following link: www.mikeselectron-icparts.com/product/spooky-tesla-spirit-radio-parts-kit/. At the time of writing, the kit costs $19.97 (£13.40) plus shipping. The kit contains all electronic parts, except for the copper wire used to make the antenna coils and a small transparent container to house the components. This can be either glass or plastic. To help identify the value of the electronic parts within the kit please see a Chap. 5 and Appendix A.

Once built, the radio works in a similar way to a crystal radio but at lower frequencies.

One interesting thing about this radio is that the diode reacts to light as well as radio waves, so varying the light conditions can have an interesting effect. The output from the radio is low as it is unpowered, so normal head-phones will not work. To hear sounds directly, a crystal earpiece must be used. These are far more sensitive than normal headphones and can be picked up online for a few dollars. Don't use a crystal earpiece with normal audio devices as this can overload the earpiece and damage it.

Software can be used such as Spectran and Spectrum Lab and set up as in Chap. 16. The output from the radio needs to be plugged into the micro-phone input on the computer. If the computer has a microphone boost func-tion, this can be used to boost the signal.

Another way that works well, especially if showing the output of the radio to a group of people, is to feed the output from the radio into an ampli-fier such as one used with an electric guitar.

Radio Astronomy's Protected Frequencies

Look at a spectrum taken from an object like a star. A number of lines will be seen; these show what elements and molecules make up the star. Radio astronomy uses radio frequencies rather than lines in a spectrum. Each fre-quency shows the existence of an element or molecule.

This is important because if we wish to look inside a nebula, visible light may be blocked by interstellar dust clouds, but the longer wavelengths of radio will travel through the dust unhindered.

Radio astronomy has a number of protected frequencies that are not used due to their importance in this endeavor. The Table 19.1 shows a list of some of these protected frequencies along with the atoms and molecules that can be studied at these frequencies.

Table 19.1 A list of protected frequencies and their corresponding atoms and molecules

Substance	Central frequency	Protected segment
Deuterium (DI)	327.3840 MHz	327.0–327.7 MHz
Hydrogen (HI)	1420.406 MHz	1370.0–1427.0 MHz
Hydroxyl radical (OH)	1612.231 MHz	1606.8–1613.8 MHz
Hydroxyl radical (OH)	1665.402 MHz	1659.8–1667.1 MHz
Hydroxyl radical (OH)	1667.359 MHz	1661.8–1669.0 MHz
Hydroxyl radical (OH)	1720.530 MHz	1714.8–1722.2 MHz
Methyladyne (CH)	3263.794 MHz	3252.9–3267.1 MHz
Methyladyne (CH)	3335.481 MHz	3324.4–3338.8 MHz
Methyladyne (CH)	3349.193 MHz	3338.0–3352.5 MHz
Formaldehyde (H2CO)	4829.660 MHz	4813.6–4834.5 MHz
Methanol (CH2OH)	6668.518 MHz	6661.8–6675.2 MHz
Ionized helium isotope (3HeII)	8665.650 MHz	8660.0–8670.0 MHz
Methanol (CH3OH)	12.178 GHz	12.17–12.19 GHz
Formaldehyde (H2CO	14.488 GHz	14.44–14.50 GHz
Cyclopropenylidene (C3H2)	18.343 GHz	18.28–18.36 GHz
Water vapour (H2O)	22.235 GHz	22.16–22.26 GHz
Ammonia (NH3)	23.723 GHz	23.64–23.74 GHz
Ammonia (NH3)	23.870 GHz	23.79–23.89 GHz
Silicon monoxide (SiO)	42.821 GHz	42.77–42.86 GHz
Silicon monoxide (SiO)	43.122 GHz	43.07–43.17 GHz
Carbon monosulphide (CS)	48.991 GHz	48.94–49.04 GHz
Deuterated formylium(DCO+)	72.039 GHz	71.96–72.11 GHz
Silicon monoxide (SiO)	86.243 GHz	86.16–86.33 GHz
Formylium (H13CO+)	86.754 GHz	86.66–86.84 GHz
Silicon monoxide (SiO)	86.243 GHz	86.16–86.33 GHz
Formylium (H13CO+)	86.754 GHz	86.66–86.84 GHz
Silicon monoxide (SiO)	86.847 GHz	86.76–86.93 GHz
Ethynyl radical (C2H)	87.300 GHz	87.21–87.39 GHz
Hydrogen cyanide (HCN)	88.632 GHz	88.34–88.72 GHz
Formylium (HCO+)	89.189 GHz	88.89–89.28 GHz
Hydrogen isocyanide (HNC)	90.664 GHz	90.57–90.76 GHz
Diazenylium (N2H)	93.174 GHz	93.07–93.27 GHz

(continued)

Table 19.1 (continued)

Substance	Central frequency	Protected segment
Carbon monosulphide (CS)	97.981 GHz	97.65–98.08 GHz
Carbon monoxide (C18O)	109.782 GHz	109.67–109.89 GHz
Carbon monoxide (13CO)	110.201 GHz	109.83–110.31 GHz
Carbon monoxide (C17O)	112.359 GHz	112.25–112.47 GHz
Carbon monoxide (CO)	115.271 GHz	114.88–115.39 GHz
Formaldehyde (H213CO)	137.450 GHz	137.31–137.59 GHz
Formaldehyde (H2CO)	140.840 GHz	140.69–140.98 GHz
Carbon monosulphide (CS)	146.969 GHz	146.82–147.12 GHz
Water vapour (H2O)	183.310 GHz	183.12–183.50 GHz
Carbon monoxide (C18O)	219.560 GHz	219.34–219.78 GHz
Carbon monoxide (13CO)	220.399 GHz	219.67–220.62 GHz
Carbon monoxide (CO)	230.538 GHz	229.77–230.77 GHz
Carbon monosulphide (CS)	244.953 GHz	244.72–245.20 GHz
Hydrogen cyanide (HCN)	265.886 GHz	265.62–266.15 GHz
Formylium (HCO+)	267.557 GHz	267.29–267.83 GHz
Hydrogen isocyanide (HNC)	271.981 GHz	271.71–272.25 GHz
Dyazenulium (N2H+)	279.511 GHz	279.23–279.79 GHz
Carbon monoxide (C18O)	312.330 GHz	329.00–329.66 GHz
Carbon monoxide (13CO)	330.587 GHz	330.25–330.92 GHz
Carbon monosulphide (CS)	342.883 GHz	342.54–343.23 GHz
Carbon monoxide (CO)	345.796 GHz	345.45–346.14GHz
Hydrogen cyanide (HCN)	354.484 GHz	354.13–354.84 GHz
Formylium (HCO+)	356.734 GHz	356.37–357.09 GHz
Dyazenulium (N2H+)	372.672 GHz	372.30–373.05 GHz
Water vapour (H2O)	380.197 GHz	379.81–380.58 GHz
Carbon monoxide (C18O)	439.088 GHz	438.64–439.53 GHz
Carbon monoxide (13CO)	440.765 GHz	440.32–441.21 GHz
Carbon monoxide (CO)	461.041 GHz	460.57–461.51 GHz
Heavy water (HDO)	464.925 GHz	464.46–465.39 GHz
Carbon (CI)	492.162 GHz	491.66–492.66 GHz
Water vapour (H218O)	547.676 GHz	547.13–548.22 GHz
Water vapour (H2O)	556.936 GHz	556.37–557.50 GHz
Ammonia (15NH3)	572.113 GHz	571.54–572.69 GHz
Ammonia (NH3)	572.498 GHz	571.92–573.07 GHz
Carbon monoxide (CO)	691.473 GHz	690.78–692.17 GHz
Hydrogen cyanide (HCN)	797.433 GHz	796.64–789.23 GHz
Formylium (HCO+)	802.653 GHz	801.85–803.85 GHz
Carbon monoxide (CO)	806.652 GHz	805.85–807.46 GHz
Carbon (CI)	809.350 GHz	808.54–810.16 GHz

The "Water Hole"

The Hydrogen (HI) or 21-cm line—the second on the list above—was the frequency chosen by Frank Drake in the 1960s as a starting point for the first attempts at SETI, as hydrogen is the most abundant element in the universe. Interestingly, the WOW! signal was found near this frequency. The Hydroxyl radical (OH), the third, fourth, fifth and sixth on the list above, was the first interstellar molecule detected using radio astronomy.

The radio frequencies between 1420 MHz and 1720 MHz are sometimes referred to as the *Water Hole*. The reason for this is due to the water molecule itself and how it is made up of the Hydrogen atom (H) and the Hydroxyl molecule (OH). When put together like this, $H + OH = H_2O$, water, hence the name the "water hole". Astronomers believe that because water is so important to every living thing on Earth, it makes sense it could be just as important to extraterrestrial life. Could this range of frequencies really be used for cosmic communication?

Useful Resources

Podcasts

Astronomy Cast produces a weekly podcast. It is produced and recorded by Fraser Cain and Dr. Pamela Gay. These podcasts can be very entertaining and well worth listening to. Here is the link: www.astronomycast.org.

Jodrell Bank Radio Observatory in the United Kingdom produces a twice-monthly podcast nicknamed "Jodcast." Here is the link: http://www.jb.man.ac.uk. Within the Jodcasts are all the up-to-the-minute advances in the field of radio astronomy, as well as what is happening within the night sky over the next month and what astronomy events to watch out for.

The website Cheap Astronomy, www.cheapastro.com, produces a podcast every Wednesday, hosted by Steve Nerlich. All of Steve's podcasts are entertaining, and all of them are done in Steve's unique style. Look through the list of past podcasts and find episodes about the Square Kilometre Array and other topics of interest.

The more humble Brains Matter website at http://www.brainsmatter.com/ has amongst the podcasts covering a wide range of subjects, several on the subject of optical astronomy and one or two on radio astronomy. These are highly recommended.

Books

Flagg, Richard S. *Listening to Jupiter: A Guide for Amateur Radio Astronomers, Second Edition.* 2000.

This book is highly recommended. It is written in a friendly, easy-to-read manner and there are several funny anecdotes, such as the one about how flames shot out of a project he had recently finished building, and the one about the night he came across an alligator. It is included as a PDF copy through a Dropbox link when you purchase the Radio Jove kit.

R. Dean Straw, Dave Hallidy, L. B. Cebik (Editors). *A.R.R.L. Antenna Book.* Newest edition available at: http://www.arrl.org/arrl-antenna-book.

This reference book covers all aspects of antennas and is handy to have lying around. It is a must-have if you decide to start designing and building your own antennas or just wish to have a greater understanding of how antennas work. The book also contains information on the suppliers of antennas and other connected items.

Romero, Renato. *Radio Nature.* Radio Society of Great Britain, 2008.

An excellent book for anyone who wishes to pursue the hobby of very low frequency (VLF) observations. Easy reading with very little mathematics. The INSPIRE project is briefly covered, as is the topic of low frequency radio emissions from the Sun. It has some good illustrations and images, including a rather strange image of a submarine's antenna..

Software

An excellent piece of software for use by the amateur radio astronomer is "Radio Eyes," available from Radio Sky Publishing. The best description of this program is that it is a radio version of an optical planetarium program, focusing on radio objects instead of optical objects. See Fig. 19.2.

This program has a clever feature, where a virtual beam pattern of a radio telescope can be set and then projected on the virtual sky of the software. Click on one of the boxes in the top right to see a list of objects within the virtual telescope's beam pattern.

The background of the program is colored to represent the different levels of radio noise from that particular part of the sky. At the default settings, a light green color indicates areas of high radio noise, while the colors purple and black show the quieter parts of the sky. Each radio source is marked on the sky with a little pink circle. Click on one of these circles; an information window pops up showing the name of the radio source and

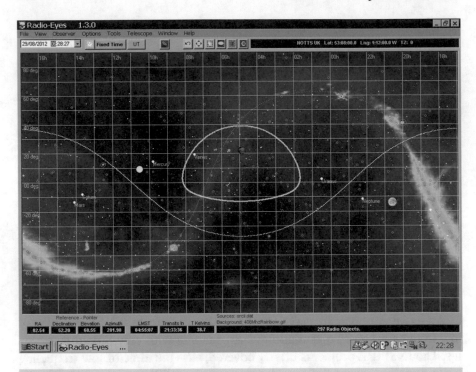

Fig. 19.2 Screenshot of Radio Eyes software

relevant information pertaining to it, such as its position, the frequency that it can be received and the strength of the radio flux measured in Janskys.

If choosing to have pulsars shown on the map, these will appear as small yellow circles. Click on one of these circles to reveal further information, such as position, frequency, spectral class and rotational speed.

The program view can be changed to show a dome view. Having both the east-west flat window and the dome view window open at the same time, you can get a three-dimensional view of the sky. As the computer cursor is moved across one window, a corresponding cursor will appear in the other window to give the exact position in the sky. This positioning can be further refined by the program's zoom feature, which is handy when trying to find an object in a particularly crowded part of the sky. Just click the mouse button and draw a rectangular mark around the part of the sky; the program will automatically zoom into this point.

You can choose to have the ecliptic, constellation patterns and Solar System objects shown on the screen. Additionally, there is a search feature. Enter the name of an object and the program will search through its database, find the radio source and point to it.

The software can be downloaded from http://radiosky.com. It comes with a 30-day free trial period. After this date the license must be purchased.

From that point forward, any updates and improvements to the program are free to download and install, therefore it is always possible to have the up-to-date version.

The website www.weaksignals.com has some useful software that may be of assistance to the more advanced radio astronomer.

Websites

- Jodrell Bank, United Kingdom http://www.jb.man.ac.uk
- Greenbank Radio Telescope, United States http://science.nrao.edu/facilites/gbt/
- Parkes Radio Telescope, Australia http://www.parkes.atnf.csiro.au
- Max Planck Observatory, Germany http://www.mpifr-bonn.mpg.de
- Arecibo Observatory, Puerto Rico http://www.naic.edu
- Royal Air Force Air Defence RADAR Museum www.radarmuseum.co.uk
- Radio Jove Project http://radiojove.gsfc.nasa.gov/
- An order form for a Radio Jove Receiver kit can be downloaded from this site.
- Radio Sky Publishing's http://radiosky.com
- From this website, you can download software such as Radio Jupiter Pro, Radio-SkyPipe and Radio Eyes, plus books such as *Listening to Jupiter*.
- British Astronomical Association (BAA) http://britastro.org/
- Search for Extra-Terrestrial Intelligence, SETI league http://www.setileague.org
- Solar and Heliospheric Observatory (SOHO) http://sohowww.nascom.nasa.gov/
- This is a good website for keeping up to date with solar activity such as sunspots and flares.
- United Kingdom Radio Astronomy Association (UKRAA) http://www.ukraa.com/
- Radio Society of Great Britain (RSGB) http://www.rsgb.org/
- This website has a number of books on the construction of antennas that may be of interest to the radio astronomer.
- The National Association for Amateur Radio (ARRL) http://www.arrl.org
- From this website, the ARRL antenna book can be ordered.
- Astrosurf www.astrosurf.com/luxorion/audiofiles.htm

From this website, all manner of sounds from space can be heard. These include pulsars, lightning discharges on Jupiter and Saturn and the sound made by the aurora at their poles. There are recordings made by the two Voyager space probes as they flew close to the Jupiter's magnetosphere.

Correction to: Radio and Radar Astronomy Projects for Beginners

Steven Arnold

Correction to:
S. Arnold, *Radio and Radar Astronomy Projects for*
Beginners, **The Patrick Moore Practical Astronomy Series,**
https://doi.org/10.1007/978-3-030-54906-0

The original version of this book was inadvertently published with few errors in Chapters 1, 2, 7 and 9. This has now been updated in the mentioned chapters in this version.

The updated online versions of these chapters can be found at
https://doi.org/10.1007/978-3-030-54906-0_1
https://doi.org/10.1007/978-3-030-54906-0_2
https://doi.org/10.1007/978-3-030-54906-0_7
https://doi.org/10.1007/978-3-030-54906-0_9

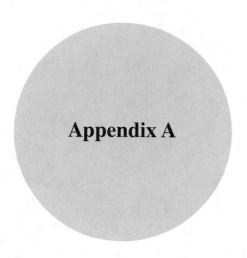

Appendix A

A common problem with electrical components is their small size. It is impractical to mark them with their value, so they can be color coded in the same way as a resistor (see Chap. 5). More commonly, they are stamped with numbers, symbols or letters to show their value—for example, a capacitor may be marked 10 μF, which means it has a value of 10 microfarads.

These tables can be used as a quick reference guide to help the reader to navigate what can sometimes seem a bit of a nightmare when it comes to identifying prefixes and symbols on electrical components.

Multiples and Submultiples Used with SI Unit

Multiples	Prefix	Symbol
10^{18}	Exa-	E
10^{15}	Peta-	P
10^{12}	Tera-	T
10^{9}	Giga-	G
10^{6}	Mega-	M
10^{3}	Kilo-	K

Submultiples	Prefix	Symbol
10^{-3}	milli-	m
10^{-6}	micro-	μ
10^{-9}	nano-	n
10^{-12}	pico-	p
10^{-15}	femto-	f
10^{-18}	atto-	a

© The Editor(s) (if applicable) and The Author(s), under exclusive license to Springer Nature Switzerland AG 2021
S. Arnold, *Radio and Radar Astronomy Projects for Beginners*, The Patrick Moore Practical Astronomy Series, https://doi.org/10.1007/978-3-030-54906-0

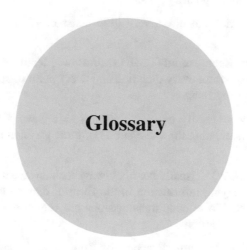

Glossary

Alternating current (AC) AC current swings between a positive value and a negative value. If seen on an oscilloscope screen, it would look like a sine wave. It has the advantage of being easy to generate and transmit along power lines.

Audio frequency (AF) Usually taken to be 20 Hz–20 KHz for humans.

Ampere/amp (A) The unit used to measure electrical current.

Amplitude The extent of the oscillation or vibration of a wave.

Amplitude Modulation (AM) This type of modulation varies the amplitude of the carrier wave to carry information.

Antenna This is the equivalent of a primary mirror or lens on an optical telescope and can take many forms. It is the collecting area for the incoming signal.

Antenna temperature A measure of the power per unit of bandwidth of a signal received by an antenna.

Bandwidth The range of frequencies or wavelengths to which an antenna is sensitive.

Blackbody A hypothetical body that absorbs all thermal radiation falling upon it and is a perfect emitter of thermal radiation.

Blackbody radiation The thermal radiation given off a blackbody. This radiation is in the form of a continuous spectrum.

Bridge rectifier A full-wave rectifier using four diodes that turns an alternating current into full-wave direct current.

© The Editor(s) (if applicable) and The Author(s), under exclusive   301
license to Springer Nature Switzerland AG 2021
S. Arnold, *Radio and Radar Astronomy Projects for Beginners*, The Patrick
Moore Practical Astronomy Series, https://doi.org/10.1007/978-3-030-54906-0

Capacitor These act like little reservoirs and store electrons. They come in many types, some of which are polarity-sensitive.

Conductor A material that offers very little resistance to the flow of electrons through it.

Cosmic Microwave Background (CMB) radiation Electromagnetic radiation that seems to emanate from every direction and every point of space; thought to be the faint echo of the Big Bang.

Carrier wave (CW) Usually but not always a sine wave. Carries information such as music by modulating the wave in one of three ways: amplitude, frequency or phase.

Continuous wave (CW) Usually but not always a sine wave that is unmodulated but is capable of carrying information in the form of dots and dashes as used in Morse code. In fact, Morse code transmissions are sometimes incorrectly called continuous wave transmissions.

Direct current (DC) Has uses in specialist welding equipment but is now only really used in low power battery-operated applications such as phones and charging devices for batteries. If DC current was seen on an oscilloscope screen, it would show a straight line.

Decibel (dB) The unit used for expressing transmission gain or loss and relative power levels.

Diode A semiconductor that will only allow current flow in one direction.

Dipole A very popular antenna used by amateurs. Usually cut to half the desired wavelength of the frequency being received.

Double sideband Similar to Amplitude Modulation, but in double sideband the carrier wave is suppressed to reduce the bandwidth.

Doppler effect The change in frequency of a wave for a stationery observer, for example the change in pitch of a siren. As the sound wave approaches, it is compressed and the pitch rises. As the siren moves away, the sound waves spread out and the pitch drops. Electromagnetic waves exhibit this property but the calculations are different from Doppler shifts in sound waves.

Electric current The movement of electrons through a conductor.

Electromagnetic Spectrum The whole sequence of electromagnetic wave energy, starting at high energy Gamma rays, through visible light to low energy radio waves.

Electron A small, naturally occurring particle with a negative charge that surrounds atoms of matter.

Electron volt The unit used to measure the kinetic energy of a particle. $1 \text{ eV} = 1.602 \times 10^{-19}$ Joules.

Farad (F)   The unit used to measure capacitance.

Ferrite collars   These are used to help stop interference entering the open end of the antenna cable and also to increase the antenna's performance. The same thing as toroid collars.

Frequency   The number of repeated wavelengths per second. Frequency is measured in Hertz (Hz), but some older books use Cycles per second (C/S).

FREQUENCY Modulation (FM)   This type of modulation varies the frequency of the carrier wave to carry information.

Frequency ranges   Each frequency has its own unique properties and uses. They are divided into different frequency bands.

Fuse   Deliberately designed to be the "eakest link" in an electrical circuit to protect the circuit from overload and damage.

Harmonic   A wave having a frequency that is an integral multiple of the fundamental frequency. For example a wave at twice the fundamental frequency would be called a second harmonic.

Henry   The unit used to measure inductance.

Hertz (Hz)   The unit used to measure of frequency.

Integrated circuit (IC)   Can contain many thousands of transistors and switches, etc. etched onto a wafer of silicon.

Io   The innermost moon of the planet Jupiter.

Io effect   Io has the effect of enhancing the radio emissions from the planet Jupiter by producing a cone of plasma that concentrates the radio waves into a beam.

Ionosphere   A number of layers in the Earth's upper atmosphere that reflect radio waves. They are affected by ultraviolet light, X-rays and other high energy radiation from the Sun and space.

Insulator   A material that highly resists the flow of electrons through it.

Interferometry   The use of two or more radio telescopes to observe an object. By combining their efforts, they increase the resolving power and sensitivity of the telescopes, relative to them being used as individual telescopes.

Isotropic antenna   A theoretical antenna used to measure the gain of a real antenna.

Jansky (Jy)   A measure of radio flux (strength) of a radio source. $1 \text{ Jy} = 10^{-26} \text{ W Hz}^{-1} \text{ m}^{-2}$. The unit was named in honor of Karl Jansky.

Kelvin   A measure of absolute temperature. The point at which no thermal or kinetic energy is present within an atom of material is known as absolute zero, or zero degrees Kelvin.

To convert degrees Kelvin to degrees Celsius: subtract 273

To convert Celsius to Fahrenheit: Degrees Celsius × 9/5 + 32 = degrees Fahrenheit

Light-emitting diode (LED)   A diode that emits light when an electrical current is passed through it. The different colors of LED's come down to the semi-conducting material they are made from. LED's are polarity sensitive and must be fitted the correct way around.

Lower sideband   A way of sending information using only the lower frequency side of a carrier wave.

Luminiferous aether   A substance that was once thought to exist throughout the universe enabling electromagnetic waves to travel in a vacuum. This was later proven wrong.

Mains power   A term used in the United Kingdom to refer to the household domestic electric power supply.

Magnetosphere   A region in space around a planet where the natural magnetic field from the planet deflects the solar wind.

Meteor   A small particle of debris or meteoroid that burns up in the atmosphere, leaving a brief streak of light, called a meteor trail, in the sky.

Meteorite   A piece of a meteor that survived being burnt up by the Earth's atmosphere and has landed on the Earth's surface.

Meteoroid   A small particle of cometary or asteroidal origin.

Micrometeorite   A very small meteorite that doesn't burn up in the Earth's atmosphere, but is quickly slowed down by it and gently drifts down to the Earth's surface.

Morse code   A way of signaling using a series of dots and dashes to represent letters, numbers and punctuation, sometimes incorrectly called Continues Wave (CW).

Ohm's Law   V = I x R. V = voltage, I = current, R = resistance.

Panspermia   A theory that life on Earth originated elsewhere and traveled to the Earth by meteorite or comet.

Perkinje effect   This effect causes some people to see some colors brighter than others even though they are the same brightness. It is well known to observers of variable stars, where the color red is often seen as fainter.

Polarization   A description of whether incoming electromagnetic waves are vertical, horizontal or circular.

Polarity   In electronics, being negative or positive.

Polarity-sensitive   Some electrical components, like resistors, can be fitted either way around, but some components, like diodes, must be fitted in the correct orientation to work properly. These are known as polarity-sensitive components.

Quadrature amplitude modulation (QAM)   This type of modulation modulates the phase of the cosine or sine waves to carry the information.

Quasars   Created from the term "QUAsi-StellAr Radio Source." These are compact objects emitting a great amount of energy over a wide range of wavelengths. All quasars have very high redshifts, which indicates they are some of the most distant and oldest objects in the universe.

Quiet Sun   The Sun when solar activity is at a minimum.

Radio frequency (RF)   This is usually taken to be 20 KHz (the upper end of human hearing) right up to 3000 GHz (the sub-millimeter part of the electromagnetic spectrum).

Radiant   A part of the sky that meteors are seen to originate from during a meteor shower. Meteor showers are named after the constellation that contains the radiant.

Radio star   A term widely used in early radio astronomy to describe any unknown radio source. Today, the term has been largely forgotten.

Receiver   The device that amplifies the incoming signal from an antenna and either increases or decreases the frequency to an audible frequency.

Rectification   The act of changing an alternating current to a direct current.

Resistor   A semiconductor used in electronics to limit current flow. The unit of resistance is the Ohm (Ω).

Semiconductor   A material that can under certain conditions act as a conductor or an insulator. Silicon is such a material.

Sudden enhancement of signal (SES)   If the ionosphere is being used to bounce a radio signal, in our case the SuperSID monitor, any changes to the ionosphere can cause the strength of the radio signal to increase or decrease when being received. If the signal strength increases, it is known as a sudden enhancement of signal.

Software-defined radio (SDR)   In its simplest form, it is an analog-to-digital converter capable of working at radio frequencies.

SI unit(s)   The International System of Units used by scientists throughout the world.

Sudden ionospheric disturbance (SID)   If the ionosphere is being used to bounce a radio signal, in our case the SuperSID monitor, any changes to the ionosphere can cause the strength of the radio signal to increase or decrease when being received. If the signal strength decreases, it is known as a sudden ionospheric disturbance of signal.

Soldering flux   Used to clean and prevent oxidization of the components during soldering.

Spectrogram   A visual graph of a sound or signal, with frequency plotted against time.

Speed of light   The speed at which light travels, equal to 300,000 km/second (186,451 miles/second).

Synchrotron radiation   Radiation from an accelerating charged particle (usually an electron) in a magnetic field.

The "Water Hole"   A range of radio frequencies between 1420 MHz to 1720 MHz. These frequencies are protected, as it is thought that if an alien civilization is indeed trying to contact the Earth, transmissions would likely be received within this frequency range.

Toroid collars   See *ferrite collars*.

Transistor   A semi-conductor that has many uses, one of which is as an amplifier.

Upper sideband   A way of sending information using only the higher frequency side of a carrier wave.

Van Allan belts   Two regions of high energy plasma in a torus shape around the planet Earth.

Vacuum   A volume completely void of matter.

Volt (V)   The unit used to measure electrical pressure.

Voltage   The pressure at which electrons are subjected to move them through a conductor.

Watt (W)   The unit used to measure electrical power. Power in Watts is calculated by
$W = I \times V$. W = Watts, I = Current, V = Voltage.

Wavelength   The physical length of a single wave pattern measured in metres.

Yagi   A multi-element antenna with high gain. These antennas are highly directional.

Zener diode   A diode with a specific reverse breakdown voltage.

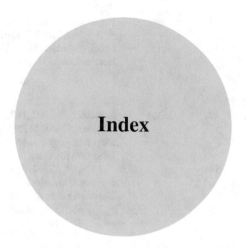

Index